# DARWINIAN AGRICULTURE

# Darwinian Agriculture

## HOW UNDERSTANDING EVOLUTION CAN IMPROVE AGRICULTURE

*R. FORD DENISON*

*PRINCETON UNIVERSITY PRESS*

*PRINCETON AND OXFORD*

Copyright © 2012 by Princeton University Press
Published by Princeton University Press, 41 William Street,
Princeton, New Jersey 08540
In the United Kingdom: Princeton University Press, 6 Oxford Street,
Woodstock, Oxfordshire OX20 1TW

press.princeton.edu

Library of Congress Cataloging-in-Publication Data
Denison, R. Ford, 1953–
Darwinian agriculture : how understanding evolution can improve agriculture /
R. Ford Denison.—Hardcover ed.
p.      cm.
Includes bibliographical references and index.
ISBN 978-0-691-13950-0 (hardcover : alk. paper)
1. Crops—Evolution. 2. Evolution (Biology) 3. Agricultural biotechnology.
4. Sustainable agriculture. I. Title.
SB106.O74D46 2012
631—dc23          2011042239

British Library Cataloging-in-Publication Data is available

This book has been composed in Sabon

Printed on acid-free paper. ∞

Printed in the United States of America

10  9  8  7  6  5  4  3  2  1

# Contents

# Illustrations

# DARWINIAN AGRICULTURE

# 1

## Repaying Darwin's Debt to Agriculture

THIS BOOK EXPLORES new approaches to improving agriculture, inspired by nature and informed by evolutionary biology. Biologists are nearly unanimous in accepting the multiple lines of evidence that life on earth has evolved and is evolving,[1-3] so applying evolutionary biology to agriculture should be no more controversial than applying chemistry and microbiology to soil science. Yet some implications of past and ongoing evolution for agriculture have often been neglected.

### Nature, Agriculture, and Evolutionary Tradeoffs

In particular, I will argue that two popular approaches to improving agriculture have tended to ignore *evolutionary tradeoffs*—that is, cases where an evolutionary change that is positive in one context is negative in another. Biotechnology advocates have often overlooked tradeoffs that arise when we genetically modify processes like photosynthesis, which have already been improved over millions of years of evolution.[4] On the other hand, people looking to nature for ideas to improve agriculture have sometimes ignored tradeoffs between the collective performance of plant and animal communities and the individual competitiveness of plants and animals. When such tradeoffs exist, evolutionary processes tend to improve individual competitiveness rather than restructure communities.[5] Therefore, the overall organization of natural communities may not be optimal, particularly as a model for agriculture. Once we drop the assumption of perfection, however, we can learn much from studying natural communities. Whether we focus on genetic improvement of crops or better management of agricultural ecosystems, identifying (and sometimes accepting) tradeoffs that constrained past evolution can often lead to new solutions to agricultural problems.

Both agricultural biotechnologists and people looking for agricultural inspiration in nature have valuable expertise and good ideas; they simply need to pay more attention to evolutionary tradeoffs. I hope this book will be read by members of both groups who want to increase their chances of success. Similarly, I hope that readers whose background is mainly in evolution or mainly in agriculture will find something here to interest them in the intersection of these fields.

I assume that readers will start—and end!—with different views on many issues, from organic farming to biotechnology (see glossary). Readers will also presumably vary in their overall familiarity with agriculture and with biology. Therefore, I include an introductory chapter on agriculture's challenges and one on evolutionary biology, emphasizing aspects key to my arguments. Many terms are defined at their first appearance in the text; I have also included an extensive glossary following the text that you can use for reference. Those who want more information, or who wonder how I have simplified a particular issue, are encouraged to read the relevant source materials listed in the references section; these are noted throughout the text using superscript numbers.

## Agricultural Challenges . . . and Two Incomplete Solutions

"Store grain everywhere," advised Chairman Mao, to ensure food security in the event of war or natural disaster. Thousands of years earlier, Egypt reportedly found a seven-year supply to be adequate.[6] More recently, in 2006, total public and private grain reserves worldwide fell to a two-month supply, as population growth outpaced increases in grain production.[7] Population growth and other trends discussed in the next chapter are predicted to increase global demand for grain (which directly and indirectly supplies most of our protein and food energy) by 40 to 60 percent over the next 30 years.[8] Although different assumptions would lead to somewhat different numbers, some increase in grain production will almost certainly be needed.

But at what environmental cost? For hundreds of years, we have increased food production by using more land and water for agriculture. Agriculture has expanded to use more water and land than any other human activity, accounting for up to 80 percent of our water use and 35 percent of the world's ice-free land surface.[9,10] Much of the remaining land is too steep, dry, wet, or cold for farming, or is set aside for parks and nature preserves. Do we really want to divert more water from rivers for irrigation, perhaps endangering fish or other wildlife? Do we really want to clear more forests or drain more wetlands to expand farmland?

I don't think so. Instead, we need to use the resources already allocated to agriculture more efficiently. For example, we need to increase the ratio of food produced to water used. This is one definition of *water-use efficiency (WUE)*. The ratio of food produced to land area used could be called *land-use efficiency*, but I will use the traditional term, *yield*. Farms account for only 3 to 5 percent of energy use in industrialized countries,[11] but rising fuel prices will make energy-use efficiency increasingly important to farmers.

*Main issues* Resource-use efficiency and food security (including quality, affordability, year-to-year reliability, and long-term sustainability) are not the only challenges facing agriculture, but they are the main focus of this book. Our goals for agriculture are considered in more detail in the next chapter.

When there are tradeoffs among multiple goals, which should have priority? Because agriculture uses a larger fraction of our water and land than it uses of fossil fuels, maybe water-use efficiency and yield should be higher priorities than energy-use efficiency. On the other hand, water and land can be reused, if we don't degrade them. Fossil fuels, once burned, are gone forever. The information in this book should help you draw your own conclusions, which may differ from mine. But even if we disagree on some answers, perhaps can we at least agree on this central question:

> How can agriculture reliably meet our needs for high-quality food and other farm products (like cotton or wool) over the long term, without environmental damage?

Two approaches that have often been proposed—rarely by the same people!—are biotechnology (such as adding genes from unrelated species to our crops, making them transgenic; see glossary)[12,13] or, alternatively, agriculture that attempts to mimic nature.[14,15] The theme of this book is that although each of these approaches has potential, both of them would benefit from greater attention to evolution, both past and ongoing.

Well-intentioned biotechnology experts may underestimate some risks of their approach. These include accidental consumption of crops grown to produce pharmaceuticals. Some less-direct risks discussed in later chapters may be even more important. Modern industrial agriculture is largely based on *monoculture*, that is, growing only one crop at a time in each field. Regionally and globally, we practice *oligoculture*, relying mainly on only a few crops, particularly corn (maize), wheat, and rice. Our major crops have been represented by many different varieties, reducing the risk that disease will destroy the crop over large enough areas to cause food shortages. Because developing each transgenic crop is so expensive, however, there are typically far fewer transgenic varieties than there are varieties developed by traditional plant breeding. If most farmers choose from only a few transgenic options, reducing overall crop diversity, are we putting too many eggs in too few baskets?

Industrial agriculture uses various methods to reduce losses to disease-causing pathogens, insect pests, and weeds, but use of toxic sprayed pesticides (see glossary) is common. The relationship between biotechnology and pesticide use is complex. The two most-common transgenic crops arguably reduce use of some pesticides, but their overall environmental impact is less clear. Widespread use of transgenic crops resistant to the weed-killing *herbicide* glyphosate presumably increases the use of that herbicide, while reducing the use of other, more-dangerous herbicides, at least until weeds evolve resistance to glyphosate. Transgenic crops with bacterial genes that make an insect-killing *insecticide* may reduce the use of insecticide sprays. But does this insect resistance in transgenic crops lead to complacency regarding other methods of pest control, such as growing different crops in sequential rotation (see glossary), increasing the risk of eventual outbreaks that would trigger greater pesticide use?

In rich countries, a large fraction of corn grain is fed to animals raised for food. Critics note the inefficiency of animal agriculture, where only a fraction of the protein and food energy (calories) in grain eaten by animals ends up in meat, milk, or eggs.[16] They suggest that we would need less grain if we ate the grain ourselves, moving "lower on the food chain." This concern predates biotechnology, but criticism of biotechnology and of other aspects of industrial agriculture may sometimes share common philosophical roots.

My own concerns about biotechnology are different. I will have more to say about the possible risks of biotechnology, but here is one of the main points I want to make in this book: *the likely near-term benefits of biotechnology have been exaggerated.* I will argue that biotechnology is unlikely to deliver soon on some key promises, such as crops that yield more grain while using much less water.

Starving research on ecologically inspired ways to improve agriculture to provide massive funding to biotechnology and its allied scientific disciplines may be fueling a biotechnology bubble. What will happen when the bubble bursts, when we finally realize that much of the money spent on biotechnology has been wasted? I hope we will then redirect some of that money to agricultural ecology and its cousins evolutionary biology, plant breeding, whole-plant physiology, soil microbiology, agronomy, and so on. But by then we may have squandered years pursuing an approach that will provide, at most, an incomplete solution to increasingly pressing agricultural problems. Population growth, depletion of natural resources, and other ongoing trends may not give us a second chance to rebalance our research priorities.

## Where Does Nature's Wisdom Lie?

Agricultural innovations inspired by nature seem more promising than many of the approaches currently being pursued by biotechnology. But we need to choose carefully which ideas from wild species and natural landscapes we apply to agriculture. How should we choose among nature's innovations?

The title of this section asks where nature's wisdom is to be found, but also whether superficial observations of nature can lead to misleading conclusions.

"Lies" are indeed common in nature. A bird may pretend to have a broken wing, to lead us away from her young. Bolas spiders eat male moths, which they lure to their deaths by mimicking the scent of a female moth.[17] But these are not the kinds of lies that worry me. We may even want to copy some of the deceptive strategies of wild plants, to mislead insect pests on our farms. Instead, I am concerned that we may sometimes mislead ourselves, if we expect to find perfection in nature. Yes, evolution has been improving nature for many millions of years. But evolution's criteria for improvement may not always coincide with our own goals for agriculture.

Leaf-cutter ants illustrate this point. I remember long lines of these ants, carrying leaf fragments back to their nest, through our rented house in Costa Rica. Our family was there because my father, later known for his pioneering research on the lichen and fern communities that cover the tops of old-growth trees,[18] took us along on a summer field trip for his students at Swarthmore College.

The ants don't eat the leaves; they use them to grow fungi and then eat the fungi. Ants have been cultivating fungi for fifty million years.[19] So if we're looking for ancient wisdom, this might seem like a good place to start. "Local food" advocates[20] might be impressed that the ants not only grow all their own food, but also rely entirely on inputs available within walking distance.

Yet the fungus farms of ants share many of the features that, in industrial agriculture, have been criticized as unsustainable. Leaf-cutter ants practice an extreme version of monoculture; each ant colony grows only one strain of one species of fungus for food.[21] Like crop monocultures grown by humans, fungal monocultures grown by ants often become infested with agricultural pests. The most harmful of these pests is another fungus, which attacks and consumes the ants' fungal crop.[22]

Like many human farmers, ants physically remove fungal "weeds" from their gardens, but they also use toxic chemicals to control the pest fungus.[23] Although these pesticides are produced by symbiotic bacteria, I will argue in chapter 11 that evolutionary aspects of this practice resemble pesticide use by human farmers more than they resemble biological control of pests by beneficial predatory insects.

Fungi are more closely related to animals than they are to plants. In other words, fungi and animals are descended from a common ancestor more recent than the one shared with plants.[24] Like animals, fungi are unable to use sunlight as an energy source, so they rely on plants for food.

The fungi cultivated by leaf-cutter ants are kept underground their entire lives, consuming leaves brought to them by the ants, much as cattle in feedlots consume grain or hay brought to them by human farmers.

Like feedlots (see glossary), ant fungus farms are inefficient in some ways. Just as meat and milk contain only a fraction of the food energy in the grain eaten by the cattle, the ants' fungal crop contains only a fraction of the food energy originally present in the leaves consumed by the fungi. If only the ants themselves could digest leaves, they might reduce their impact on the environment by harvesting fewer leaves and consuming them directly, thereby eating "lower on the food chain."

To summarize, leaf-cutter ants practice monoculture, use pesticides, and manage inefficient fungi as if the fungi were cows in crowded feedlots rather than in pleasant pastures. Ants have been following these practices for fifty million years.

As we look to nature as a source of ideas for agriculture, how should we react to this information? We have at least three options.

First, we could continue to insist that nature is perfect, but deny those aspects of nature that are inconsistent with our ideals. This is a very popular approach, but not one I advocate.

Second, if we believe that agriculture should copy nature whenever possible, we could endorse monoculture, pesticides, and feedlots, without any reservations. For example, one biotechnology advocate has argued that it is acceptable for us to use toxic pesticides, because many plants use toxic chemicals to defend themselves from insect pests.[25] But I don't like this mindless-mimicry-of-nature option much either.

Or, third, we could choose carefully *which* ideas from nature we apply to agriculture. If some of the "wisdom" of the leaf-cutter ants turns out to be "lies," how can we avoid being misled in other cases, where the risks of mimicking nature are less obvious? How can we be sure we are copying only nature's best ideas?

In 2009, we celebrated the 200th anniversary of Charles Darwin's birth and the 150th anniversary of his best-known book, *The Origin of Species*. Darwin saw agriculture as a rich source of information for understanding nature, an approach that, he complained, was often "neglected by naturalists."[26] His best argument for the power of natural selection—the central idea in his book—was the success of plant and animal breeders, greatly improving crops and livestock simply by selecting which individual plants and animals get to reproduce. In borrowing this key idea from agriculture, Darwin incurred an intellectual debt, acknowledged by him and inherited by today's evolutionary biologists. Can evolutionary biology repay Darwin's debt to agriculture in the same currency of ideas, identifying evolutionary innovations in the natural world that we can adapt to agriculture? If so, where in the natural world will we find these innovations?

To answer this question, we need to determine which aspects of nature have been improved most by evolutionary processes. I will argue that evolution has improved trees much more consistently than it has improved forests. In other words, nature's wisdom is to be found more in the adaptations of individual plants and animals than in the overall organization of the natural communities and ecosystems (see glossary) where they live. Often, individual adaptations that have been tested by millions of years of evolution will be more sophisticated than anything biotechnologists can imagine and implement. For example, evolution is unlikely to have missed simple, tradeoff-free opportunities to improve biochemical processes like photosynthesis.[4]

But tradeoffs that constrained past evolution need not always limit us today. Tradeoffs between adaptation to past versus present conditions suggest various options for crop improvement through traditional breeding methods or biotechnology.[27] Tradeoffs between individual competitiveness and the collective performance of plant and animal communities may be even more important.[28] For example, although cooperation between species is already common in nature, an evolutionary perspective suggests considerable room for improvement.[29]

When evolution has already been working on a problem for millions of years (improving drought resistance, for example), keeping or copying nature's innovations will often be our best option. But when past evolution has not been fully consistent with our goals, we may be able to improve on nature. Often, this will involve accepting tradeoffs previously rejected by evolution.

## Overview of This Book

Here is a brief overview of this book. The next two chapters introduce agriculture and evolution, respectively, but even those familiar with these areas may find new information or ideas to consider. Chapter 2 will discuss some of the challenges that agriculture is facing now or will face soon. Chapter 3 will review some definitions and concepts from those aspects of evolutionary biology that are central to subsequent arguments.

Chapter 4 proposes three core principles that will be developed throughout the rest of the book. First, natural selection is fast enough, and has been improving plants and animals for long enough, that it has left few simple, tradeoff-free opportunities for further improvement. Therefore, implicit or explicit acceptance of tradeoffs has been and will be key to crop genetic improvement, through biotechnology or traditional breeding methods. Some tradeoffs, such as adaptation to conditions that no longer exist, will be easier to accept than others. Second, nature's testing of natural ecosystems merely by endurance is weaker than the repeated competitive testing of individual adaptations by natural selection. Testing by endurance shows sustainability—some natural ecosystems have persisted for millennia—but there may still be considerable room for improvement. We can use what we learn about natural ecosystems to design better agricultural ecosystems, but simply copying the organization of natural ecosystems is unlikely to improve the performance of our farms, by most criteria. Last, I advocate a greater diversity of crops—not necessarily in mixtures—and a greater diversity of research approaches, to hedge our bets against future uncertainty.

Chapter 5 builds on chapters 3 and 4 to argue that some of biotechnology's stated goals, such as more efficient use of water by crops, are unlikely to be achieved without tradeoffs. Possible benefits and risks from biotechnology are discussed. Chapter 6 explores natural selection's limitations: it has bequeathed many sophisticated adaptations to individual plants and animals, but it has not consistently improved the overall organization of the natural communities where they live. The available evidence suggests that no other natural process has optimized natural communities either. Building on these conclusions, chapter 7 evaluates some of the more-popular proposals for how agriculture might attempt to mimic nature. In each case, I suggest some reasons for caution.

Beginning in chapter 8, I turn from criticizing popular but problematic approaches and take a more optimistic view, describing past successes and future opportunities. Many past agricultural improvements have involved accepting tradeoffs previously rejected by evolution, reversing some negative effects of past natural selection. For example, humans have selected for greater cooperation among plants, improving the collective performance of crop-plant communities by sacrificing some individual-plant competitiveness. Selection for more-cooperative plants has not usually been deliberate, but it can be. Chapter 9 focuses on cooperation between two species. Such cooperation is already widespread, but there is plenty of room for improvement. Understanding tradeoffs between the interests of symbiotic partners is key to unlocking this potential.

These first nine chapters mainly emphasize implications of past evolution. Chapter 10 considers ongoing evolution, particularly as it relates to control of agricultural pests. Chapter 11 discusses fungus-growing ants in more detail and extends our search for nature's wisdom to interactions among more species. I argue that natural landscapes need not have optimal structure to be valuable sources of ideas. Last, chapter 12 summarizes key conclusions and cautions against exclusive reliance on any single approach, even those proposed in this book. I argue that although processes similar to competitive natural selection may help us choose the best ideas, we should also hedge our bets by maintaining a diversity of approaches.

# 2

## What Do We Need from Agriculture?

IF OUR GOAL IS TO IMPROVE AGRICULTURE, what do we want to improve? Some important criteria include productivity (yield per acre, to use no more land than necessary), efficiency in the use of scarce resources (to use no more water than necessary, for example), stability over years (to prevent even occasional famines), and sustainability (to maintain all of these benefits over the long term). Improvements in any of these will affect the billions of us who live in cities, both through effects on our food supply and through effects on the availability of land and water for other uses. Other important goals include the health of wildlife living on or near farms and the welfare of people who work on or near farms.

### Agriculture Affects Everyone, Not Just Farmers

My brother Tom is a successful farmer. Shunning synthetic chemicals, he uses only *organic* methods. His family's small farm has been their main source of income for many years. Their fruits and vegetables are locally renowned for their quality. One time while I was visiting, he donated a whole acre of melons (a new variety he was testing) to charity. "These aren't any better than what people can buy at the store," he explained, "I can't put the Denison Farms label on them." His golden raspberries, his crunchy-sweet persimmons, and even his carrots are amazingly tasty. Someone as smart and hardworking as he is could certainly make more money doing something else, but there's more to life than money. He has a wonderful family, worthwhile work, loyal customers, and a great view of the Oregon Coast Range from his workplace.

When my grandfather was young, most Americans were farmers. Now, only a few of us are. So although I will sometimes use examples from Tom's farm, my focus in this chapter is on what the rest of us, nonfarmers, want and need from agriculture.

Here's an example of the difference. The Lundberg family of California has been growing rice for generations. Much of the organic rice sold in the United States comes from their farms, as does some conventionally grown rice. Years ago, I heard a talk by one of the Lundberg brothers. At the time, they could sell a pound of organic rice for twice what they got for a pound of rice grown with conventional fertilizers and pesticides. So, he said, if they could consistently get half the yield (pounds of rice per acre), they would grow only organic rice. After they first stopped using herbicides on a given field, they usually got good yields for the first few years. But certain weeds built up over years, eventually reducing yields to less than half what they could get with herbicides. His talk discussed various innovative methods they were using to try to control weeds in their organic fields. They may have solved this problem by now. But this sort of dilemma is common, with implications for consumers, not just farmers. What should nonfarmers think of farmers producing half as much rice at twice the price? To put this another way, if farmers use twice as much land to grow the same amount of rice, should we pay them twice as much?

I should point out that there are also situations where organic methods give higher yields than conventional methods. In my own research, for example, this was sometimes true for organic tomatoes, particularly when spring weather was unusually wet.[30] But what about those cases where organic yields are consistently lower? Should we encourage farmers to use organic methods, even if that means we (including those of us struggling economically) have to pay higher prices, and even if it means using more land for agriculture?

Decisions made by individual farmers affect the rest of us, and vice versa. For a farmer, growing half as much rice and getting twice the price per pound may be acceptable. Given the health risks of pesticides to farmers, they might reasonably prefer the low-yield/high-price option. But what about the rest of us? Is it worth paying twice as much for rice grown without pesticides? Individual consumers might give different answers, depending on their particular health concerns and what fraction of their household budget is spent on rice.

Choices made by individual farmers add up in ways that affect consumers. Similarly, choices made by individual consumers add up in ways that affect everyone. If consumer demand for higher-priced organic rice increases, so that conventional farmers convert more of their land to organic production, rivers downstream from rice-growing areas will presumably have less pesticide pollution. Beneficiaries could include people who catch fish in those rivers and people whose cities draw drinking water from those rivers, whether or not those individuals buy organic rice.

Individual farmers and individual consumers may use different criteria in deciding whether to grow, or buy, organic crops. But agricultural practices are also influenced and constrained by public policy, ranging from laws restricting pesticide use to government funding for agricultural research. So what does the public need and want from agriculture?

## Reliable Food Surpluses Are History

First and foremost, we depend on agriculture for most of our food. Growing enough food for everyone is no longer something we can take for granted. Since 2003, world grain stocks have never exceeded 80 days' consumption.[7] In 2011, world food prices set a new record.[31] Year-to-year variation in food supply is affected by many factors, such as droughts. On average, however, recent increases in world food production have not kept pace with recent increases in world population.

*Scary* [handwritten annotation in left margin]

The relationship between population growth and food supply has been controversial at least since Reverend Thomas Malthus published *An Essay on the Principle of Population* in 1807. Malthus argued that human population will grow geometrically, unless it is controlled somehow—he suggested delaying marriage to decrease birth rates. He used the United States as his example, "where the means of subsistence have been more ample, the manners of the people more pure, and consequently the checks to early marriages fewer, than in any of the modern states of Europe, the population has been found to double itself in twenty-five years."[32] At that rate, which he considered close to the maximum possible, U.S. population would have doubled eight times, increasing 256-fold, in the 200 years since those words were written. The actual increase has been only about 48-fold, some of it due to immigration. World population has increased about 7-fold. There are, however, many examples of populations doubling in 25 years. Slower average growth results from various factors, from birth control to famine to war.

*did not* [handwritten annotation in left margin]

Malthus's claim that food supply can grow only arithmetically is on less firm ground than the potential for exponential population growth. There is no obvious reason why food supply should follow a consistent trend over decades, whether arithmetic or geometric. Malthus was rather generous in his linear trend, however. He assumed that, at most, food production in the UK "might be increased every twenty-five years, by a quantity of subsistence equal to what it at present produces." At this linear rate, food production would have increased 8-fold in 200 years, slightly more than the actual increase in world population. In fact, however, wheat yields in England increased only 4.7-fold, from about 1.5 to 7 tons/hectare.[33] So Malthus, widely viewed as a pessimist, actually overestimated the potential for yield increases. Nonetheless, data for shorter periods when yields increased faster than average are often used to claim that technological advances have proved Malthus wrong.

World grain production per person peaked around 1984. Since then, population growth has outpaced increases in production.[7] By 2006, worldwide grain production per person had fallen to 1.8 pounds (0.83 kilogram) per day. If none of this grain were spoiled, eaten by rats or

*For Make world grain production info public + transparent* [handwritten annotation at bottom]

farm animals, or fermented into ethanol, then it would provide more than enough protein and energy (3000 calories per day) for a healthy diet. However, the efficiency of conversion from grain calories to meat calories (chicken or pork) is only 15 to 25 percent,[16] so 1.8 pounds of grain would yield less than 1000 meat calories per person per day. In other words, the world currently produces enough food for an adequate grain-based diet for everyone, but not enough for everyone to eat a meat-based diet.

Still, it might seem that there would be no need for further increases in grain production, if only (1) population growth ceased immediately and (2) meat consumption decreased greatly. Neither of these is likely to happen, however. First, there are far more people of child-bearing age or younger than there are people dying of old age. Therefore, even an immediate and universal switch to two-child families would take decades to slow and stop population growth. Second, many people like to eat meat. As people who could rarely afford meat in the past become richer, global meat consumption is likely to increase.

Population growth is not the only trend that may cause food demand to outpace supply, but it merits some additional discussion. Experts agree that further increases in human population are inevitable, but they disagree about how much growth we should expect. Often, birth rates have fallen as incomes or educational levels have increased. If birth rates in every nation, ethnic group, and subculture all fall to replacement levels, then population growth will cease, eventually.

But what if every country maintains its current rate of population growth, low in some countries and high in others? Then high-growth countries will become a larger fraction of the total each year, while low-growth countries will have less and less effect on total world population growth. Under this hypothetical scenario, population expert Joel Cohen has estimated that world population in 2100 could reach 109 billion,[34] an 18-fold increase from 2000, as high-growth countries increasingly dominate world population.

Most estimates, including those from the same author,[34] are much less than that, but some population growth is generally considered inevitable. One expert opinion is that population growth and economic growth (leading to more meat consumption) will require a 40 to 60 percent increase in yields of major grain crops between 2000 and 2030.[8] It remains to be seen whether such an increase is possible at an acceptable environmental cost.

An additional strain on food supplies will be increased industrial uses for crops, competing with their use for food. Recent increases in fermentation of corn into ethanol in the United States could perhaps be reversed, despite the political strength of the ethanol industry. But other industrial uses of crops are almost certain to increase, and they probably should. Ink made from soybean oil is just one example. Lubricants, plastics, and adhesives are mostly made from oil or natural gas today. As fossil fuels become increasingly expensive, these materials will be made from crops. The lubricant on a bicycle chain is such a small part of the cost of the chain that we would readily pay ten times what we do today for lubricant, rather than have the chain wear out prematurely.

The more profitable it becomes to grow crops as chemical feedstocks, the more land will be taken out of food production. Some consumers may not want to eat transgenic crops, but who asks whether the plastic in a cell phone was made from soybeans that were genetically engineered to produce a vegetable oil more suited to plastic production? Increasing human population, likely increases in demand for meat, and conversion of crops to ethanol or other industrial products mean that current world crop production will probably not be enough to meet future needs.

## Transportation Is an Additional Problem

The places that grow most of the world's food are not the same places where the most food is needed. This creates several related challenges. Even if global food supply per person is adequate today, this does not guarantee that the poorest people in the poorest countries get enough food. If some people want to (and can afford to) consume more than they need, others may get less than they need. If you believe that equal distribution of food worldwide is actually likely to happen soon, then your estimate of how much food the world needs will be lower. If you expect inequality among and within countries to persist for some time, however, then we will need to produce more food.

Transportation of food itself is not the main problem. An adult can carry enough grain to meet his or her calorie and protein needs for a month. With even a simple cart or dugout canoe, he or she can transport enough for a large family. Over large distances, transport by ship, barge, or rail is so energy-efficient that it adds little to the energy cost of food. A consumer who drives several miles to a farmers' market just to buy a few local tomatoes may use more fuel, per tomato, than a truck full of tomatoes, traveling across the country. It has been claimed that the difference in energy needed to raise lambs in the UK versus New Zealand is greater than the energy cost of shipping meat from New Zealand.[35] Within the United States, our food system's contribution to the *greenhouse gases* that contribute to our warming climate comes mostly from crop production and only secondarily from transportation. Only 4 percent of agriculture's total comes from farm-to-store transportation.[36]

Agriculture faces more serious transportation problems, however—particularly the transport of animal manure. The problem is manure's high weight, relative to the value of the plant nutrients it contains. In contrast, the economic and energy costs of transporting wheat grain are small relative to its value to bakeries and consumers in cities. But every metric ton (1000 kilograms) of wheat leaving a farm carries with it 23 kilograms of nitrogen and 4 kilograms each of phosphorus and potassium.[16] These and other nutrients removed from the farm in the harvested grain must somehow be replaced, perhaps by applying manure to the soil.[37] Otherwise, soil fertility will gradually decrease, reducing future crop yields.

This process can take some time. When a friend e-mailed me that he had been "farming sustainably for three years," I wondered how he could already be sure that his methods were sustainable (see glossary) after only three years. Remember the gradual buildup of weeds in organic rice fields? In my own research, we harvested wheat from unfertilized control plots for nine years, before the downward yield trend became statistically significant.[30] One experimental organic farm managed without external inputs (except for atmospheric nitrogen taken up by symbiotic bacteria) for almost 30 years, but eventually they had to start importing nutrients, in the form of manure from other farms.[38]

So, even if farm-to-market transportation isn't a big problem, what about the return trip? Can we return the nutrients from cities to farms? Some of the nutrients end up in our bodies and aren't available as fertilizer until after we die, if then. Much of the rest is in garbage or sewage, where potentially valuable nutrients are mixed with bacteria, viruses, various toxins (including residues from legal and illegal drugs), and water. Water is a problem because it makes these waste products heavy, and therefore expensive to transport, relative to the value of their nutrient content.

*[margin note:]* Composting

Averaged over the planet, both nitrogen and phosphorus are applied at rates that exceed their removal in harvested crops, an excess that contributes to pollution. Yet 30 percent of farmland receives too little phosphorus,[39] and nitrogen shortages are common, especially in Africa.[40] There are many reasons for this distribution problem, but the energy cost of transporting manure is an important factor.

Industrial production of meat and milk makes the nutrient transportation problem even worse. In the United States, for example, many chickens eaten in New York are raised in Alabama, using grain grown in Nebraska. Ideally, the chicken manure would be returned to Nebraska farms, replacing a large fraction of the nutrients exported in the harvested grain, while reducing pollution problems in Alabama. But the value of that manure to those farms may be much less than the cost of transporting it such a long distance.

*[handwritten note:]* Won't work     smaller, more local farms

This is true for dollar value, but it can be true even if our main concern is saving energy. I asked students in my crop ecology class to calculate the energy cost (per pound of nitrogen) of hauling wet manure compared to producing nitrogen fertilizer from natural gas and then transporting the fertilizer across the country. Because the synthetic fertilizer has so much more nitrogen per pound, relative to manure, it actually uses less energy to make and transport the fertilizer across the country than it does to haul manure for more than a few miles.

There are at least three possible ways to meet the long-term nutrient needs of crops, while reducing water-pollution problems caused by excess nutrients in animal feedlots and in cities. First, we could accept one-way transfers of nutrients from farms to cities. We would then work to increase nutrient supply to crops and decrease water pollution from cities, as two separate problems. This is the main approach used now. Second, we may be able to invent practical and efficient ways to remove *pathogens* (microbes that cause disease), toxins, and especially water from sewage, garbage, and animal manure, so that these cleaner and lighter waste products can be transported long distances back to farms at a reasonable cost.

Third, we could perhaps encourage migration from cities to rural areas, reducing the cost of returning nutrients to farms by shortening the distance. But this approach could create other problems. For example, deforestation is more correlated with rural population size than with total population[41]—people in cities may use wood from distant forests, but they don't cut them down for fuel or gardens. A dispersed population might decrease transport costs for food and manure, but it would make mass-transit less practical. That would increase transport costs for people traveling to jobs, universities, or cultural and sporting events, assuming that such amenities could persist without cities.

As long as we do have cities, net transfer of nutrients from farms to cities in harvested farm products is likely to be a significant problem. Potential decreases in soil fertility caused by these transfers exemplify a broader challenge facing agriculture: the challenge of long-term sustainability.

## Sustainability Is Essential

There are serious concerns about whether even current levels of production can be maintained—that is, whether current agricultural practices and land-use trends are sustainable. Furthermore, food production varies from year to year. Even if we produce enough food to feed everyone in an average year, what about years with unusually bad weather, or the inevitable crop-disease epidemics?

Let's consider sustainability first. A standard definition of *sustainable* is "possible to continue indefinitely." A broader definition would be "meeting current needs in ways that preserve options for future generations." I prefer these results-focused definitions to lists of specific practices, whose sustainability may turn out to be less than we hope.

If an agricultural practice has unacceptable short-term consequences—we may argue about what is acceptable—then we don't need to ask whether it could continue indefinitely.

Sustainability becomes an issue only when there may be serious long-term problems that are not seen in the short-term. For example, burning oil releases various pollutants that may have immediate effects on human health. But what about longer-term effects? Burning oil also releases *carbon dioxide*, a greenhouse gas with long-term effects on climate. Also, oil used for a shopping trip today will not be available to fuel an ambulance in the future. It is these longer-term consequences that raise questions of sustainability.

Some examples of things that obviously cannot continue indefinitely are harvesting natural resources (such as fish, trees, or fresh water) faster than they are replenished, excessive borrowing by governments or individuals, underinvestment in maintaining physical infrastructure or in education, and birth rates higher or lower than death rates. All of these can continue for decades (and often have), but not indefinitely.

There are two kinds of threat to agricultural sustainability. First, some farming methods can cause crop yields in a given field to decrease over decades. That is, the farming practices themselves are intrinsically unsustainable. Second, some resources that are key to current farming methods are being used faster than they can be replenished by natural processes. Even when agriculture is not the main user of a particular resource, depletion by other users can cause shortages that affect agriculture.

Soil loss due to erosion and excess accumulation of salt in soil are two examples of intrinsic threats to sustainability. Rates of soil loss often exceed the slow rates of replacement by natural soil-forming processes, especially when soil is disturbed by plowing or cultivation to kill weeds. Irrigation water always contains some salt, which can accumulate to levels that harm plants, unless it is removed via natural or artificial drainage.

These two examples of threats to sustainability illustrate two important points. First, only one of these problems is caused by excessive chemical inputs; furthermore, the input of concern is salt naturally present in irrigation water, rather than some synthetic chemical. So, although reducing synthetic inputs like pesticides may often be a good idea, that may not be enough to guarantee sustainability. Second, both erosion and salt accumulation can occur on either organic or conventional farms. If killing weeds with herbicides allows conventional farmers to use less tillage, then we need to compare the environmental impact of herbicides with possible increases in erosion from tillage.

My brother follows organic rules, but is he farming sustainably? His farm is reasonably flat; I have never seen any sign of soil erosion there. Salt accumulation mostly seems to be a problem in hot, dry climates, where farmers apply much more irrigation water than Tom needs in Oregon's relatively cool summers. In any case, rainy winters and reasonably good drainage probably remove any salt added in irrigation water. As far as I can tell, Tom's farming practices are maintaining or improving the soil over time. If one of his children chooses to farm, the farm should be in good shape when he retires. So he does seem to be farming sustainably.

The same is probably true of many farms, including some that use "conventional" (as opposed to organic) methods. But we should take a longer view. Could gradual changes, too slow to be noticed by individual farmers, be disastrous over decades or centuries? Absolutely.

One example comes from Mesopotamia, currently known as Iraq. Five thousand years ago, wheat and barley were equally important there. The evidence comes from counting impressions made by wheat and barley seeds in pottery recovered from archaeological sites.[42] By about 4000 years ago, however, only barley was being grown. Why did farmers switch to barley, when wheat is usually preferred? Barley is more tolerant of salty soils than wheat is, so the switch was probably driven by increasingly salty soils. Ancient documents show that even barley yields eventually decreased. Irrigation, without enough drainage to remove salt added in irrigation water, is the most likely culprit.

No individual farmer would notice a trend that took centuries to cut yields in half. Long-term trends can be obscured by year-to-year variability in weather. Stories of higher yields in past generations might be considered myth, like stories of giants, rather than fact. Even if a downward yield trend were recognized—salt accumulation doesn't always take that long—causes and solutions might not have been obvious. If gradual trends are hard to detect on working farms, what about farms dedicated to research? The prime example is the Rothamsted Agricultural Research Station, north of London. My colleague Dennis Bryant and I visited Rothamsted in 1993. We were starting a long-term experiment at the University of California's Davis campus (UC Davis), focusing on sustainability issues in irrigated agriculture. We hoped to learn from Rothamsted's 150-year history of long-term experiments.

We toured the long-term wheat plots on a beautiful day in late May. Most plots had few weeds, because they had been sprayed with herbicides. Some of the most interesting differences, however, involved various herbicide-free control plots. Those that had been supplied with either synthetic fertilizers or cow manure—for 150 years—had a variety of weeds mixed with the wheat, including beautiful red poppies in bloom. No-herbicide plots that had received no nitrogen, however, were dominated by a single weed species, vetch. Although vetch is considered a weed in wheat, at UC Davis we grew it as a green manure (see glossary), plowed in to improve the soil. Vetch doesn't need nitrogen fertilizer, because its symbiotic root-nodule bacteria (*rhizobia*) convert nitrogen from the atmosphere to forms plants can use. We grew vetch during Davis's mild winters, when it would use little water, then plowed it under in the spring, making much of its nitrogen available to subsequent crops.[43]

Rothamsted was the home of John Lawes, inventor of a process for making high-phosphate fertilizer. Back in 1843, most farmers used organic fertilizers, especially animal manure. Sodium nitrate mined in Chile was becoming popular, however. Chemists define organic molecules as those that contain carbon. Sodium nitrate contains only sodium, nitrogen, and oxygen, so it is inorganic, by this definition. Some organic rules have allowed limited use of mined sodium nitrate, however, because it is considered "natural."

Sodium nitrate and inorganic phosphate fertilizer were readily available sources of nitrogen and phosphorus, and they gave good crop yields, at least for a few years. There were rational concerns about their sustainability, however. Animal manure supplies other plant nutrients in addition to nitrogen and phosphorus. Also, while some of the nitrogen in manure is already in inorganic forms (ammonium or nitrate), manure also contains various organic molecules that release available nitrogen slowly. Some organic molecules in manure break down so slowly that they can last thousands of years in soil. These long-lasting organic molecules from manure and other sources contribute to *soil organic matter*, which helps retain various plant nutrients in the soil, increases the soil's ability to store water, and provides other benefits. Given all these benefits, are organic fertilizers like manure essential for sustainability, or can inorganic nutrients like sodium nitrate and phosphate fertilizer meet long-term crop needs?

At Rothamsted, Lawes and his collaborator Joseph Gilbert decided to find out. In 1843, they started a series of experiments comparing crops grown with animal manure to crops grown with various combinations of inorganic fertilizers. Some of these experiments, including the wheat experiment we visited in 1993, ran much longer than planned. As expected, adding manure for 150 years led to higher levels of soil organic matter, relative to 150 years of inorganic fertilizer. In fact, soil organic matter doubled with manure addition, although this took 75 years. But soil organic matter in the inorganic fertilizer plots stayed reasonably high also. Apparently, plowing nonharvested parts of the crop into the soil each year added enough organic matter to maintain essential soil functions. Wheat yields in plots with either manure or inorganic fertilizer have been similar over 150 years, and both have increased over decades.[44]

In this section, we are focusing on threats to sustainability and particularly on intrinsically unsustainable farming practices, those that lead to long-term decreases in crop yield. By this criterion, can reliance on inorganic fertilizers ever be sustainable? A century and a half of increasing wheat yields with inorganic fertilizer is certainly encouraging. But maybe that's not enough time to see some problems that might develop gradually over centuries. Remember the thousand-year decrease in wheat yield in Mesopotamia?

Furthermore, the importance of the organic matter from manure might be greater at other locations. At another long-term research site, also managed by scientists at Rothamsted, yields with inorganic fertilizers have been lower than those with manure.[45] The difference, apparently, is the relatively high clay content at Rothamsted relative to the other site. Clay, which consists of the finest soil particles, substitutes for some of the beneficial effects of soil organic matter from manure, particularly improving storage of water and nutrients in soil.

Not all of the agricultural practices tested at Rothamsted proved sustainable. In a four-crop rotation (see glossary), turnip yields gradually increased over the first 40 years. Based on this trend, the system appeared to be sustainable. But then yields slowly started to decline. After another 40 years, yields approached zero, and this experiment was abandoned. These yield declines were mysterious at first, but they were eventually explained, using soil samples collected over many decades.

When we visited Rothamsted, Johnny Johnston, who knows more about the long-term experiments at Rothamsted than anyone, took me to see their archival sample collection. We passed by the sprawling brick Rothamsted Manor, now used as housing for visiting students and scientists from around the world, and entered a simple concrete building. There we saw row upon row of towering shelves, covered with a variety of sample containers. A glass jar, corked and sealed with lead foil, held a tissue sample from a pig slaughtered over a century ago. Samples of grain, harvested over decades, contained enough DNA from fungi infecting the crop that scientists were able to link past crop-disease epidemics to trends in air pollution over 160 years.[46] But we were looking for soil samples.

When the first soil samples were collected from the four-crop rotation at Rothamsted, the science of soil chemistry was in its infancy. In particular, our modern measure of acidity, pH, had not yet been defined. Fortunately, Rothamsted scientists collected and saved soil samples anyway. Modern analyses of these soil samples revealed a trend of increasing soil acidity (decreasing pH) over past decades. It is now known that some nitrogen fertilizers can acidify soil. The soil never got acid enough to hurt the crops directly, and at first there was little apparent effect. Remember that yields actually increased for the first 40 years. But eventually the soil became acid enough that a particular acid-loving fungus became established. It was this fungus, a serious pathogen of turnips, that caused the eventual declines in yield.[47] The acidification could have been reversed by adding lime. But by the time the problem was understood, it was too late. The fungus was too well established.

This was the sort of problem we were trying to prevent in California, with our own long-term experiment at UC Davis: problems that might be overlooked until it is too late to do anything about them. I directed the first 10 years of the Long-Term Research on Agricultural Systems (LTRAS) project, as a crop ecology professor in the Agronomy and Range Science department at UC Davis. LTRAS was designed as an early warning system to measure and analyze long-term threats to agricultural sustainability, especially those that are too slow to be detected and studied on ordinary farms. Sixty of the 72 one-acre plots at LTRAS were dedicated to a 100-year comparison of ten conventional, organic, and alternative cropping systems. The other one-acre plots, and some smaller ones around the edges, have been used for various shorter-term experiments.

Inspired by Rothamsted, we made collection and preservation of archival samples an integral part of LTRAS. Among the gradual trends we hoped to study were evolutionary changes in soil microbes and weed populations. (The oldest experiments at Rothamsted started 16 years before Darwin's *Origin of Species* was published, but Ronald Fisher made major contributions to our understanding of both evolution and statistics while working there.[48, 49]) At LTRAS, Professor Robert Norris set up four "weed preserves" around the site. These were small plots where weeds were allowed to complete their life cycle and produce seeds. Although Norris himself was nearing retirement, he collected seeds from each common weed species and stored them for research by future generations of scientists. If pigweed populations evolve different rooting patterns in irrigated versus nonirrigated plots, researchers will be able to compare DNA from these two evolved populations with their common ancestor, preserved in the Norris seed collection.

Similarly, I collected and freeze-dried some soil samples (in addition to our regular dried soil samples), using surgical gloves and flame-sterilized tools to prevent contamination of the soil microbe populations with bacteria from my hands. I wanted to preserve a "time-zero" record of soil microbial populations, for comparison with future changes. My idea was that someday it would be possible to characterize microbial communities by counting species-specific genes. This became possible sooner than I expected. Yutaka Okano, a UC Davis student from Japan, was among the first people to count particular groups of soil bacteria based on their DNA, using soil samples from LTRAS.[50] As an example of the sorts of trends we hoped to detect, we expected differences to develop in the relative abundance of fungi in organic versus conventional plots. In fact, this was already evident a few years into the experiment: mushrooms came up in the three replicate plots of one of the organic treatments, but not elsewhere.

There was some initial opposition to LTRAS, from people with strong opinions about sustainability, but they apparently calmed down after a few years. Perhaps, like certain fictional philosophers,[51] it took critics of LTRAS a while to realize that a 100-year experiment might stimulate public interest in their opinions, while providing conclusive results only after they were dead. One organic-farming activist who visited LTRAS was apparently convinced that our research was somehow controlled by chemical companies, until he decided that we had "the best green-manure crops in the county."

Those green-manure crops were mixtures of peas and vetch. As mentioned earlier, we grew them to add organic matter and nitrogen to the soil. Much of this nitrogen came from rhizobia in their root nodules. In LTRAS's early years, wheat that followed green manure had yields similar to fertilized wheat.[30] This could be seen as evidence that the green manures were meeting wheat's nitrogen needs, making them a sustainable alternative to nitrogen fertilizer. But that conclusion would have been premature. We know that because we also included nonfertilized controls. In the early years, wheat yields in some of these unfertilized plots also had yields similar to the fertilized plots. Soil organic matter typically contains enough nitrogen, released gradually, to meet crop needs for years. So LTRAS needs to continue the wheat treatments at least until the unfertilized controls run out of soil nitrogen, before drawing conclusions about the sustainability of green manures as a nitrogen source for wheat.

LTRAS and other long-term experiments can study only direct threats to agricultural sustainability, such as nutrient depletion, erosion, salt accumulation, or soil acidification. What about the wider problem of resource availability? Only 3 to 5 percent of our energy consumption is on farms or in fertilizer factories.[16] This fraction increases if we include everything related to food transport and processing, including driving to the grocery store, refrigeration, and cooking. Food is more essential than air travel or air conditioning, so agriculture will probably compete effectively for fossil fuels, even after they become scarce.

You may have heard that "when the oil runs out, we will have to stop using synthetic fertilizers and pesticides." For better or worse, this is not true. The inputs used for nitrogen fertilizer production today are nitrogen gas (an inexhaustible 80 percent of the atmosphere) and natural gas.

Natural gas will become increasingly expensive as reserves are depleted. However, we can substitute hydrogen gas, which can be made by running electricity from renewable sources through water. For example, the University of Minnesota has a wind-powered nitrogen-fertilizer factory under construction.[52] Similarly, synthetic pesticides can use plant products as raw materials, instead of oil. So higher fuel prices will discourage overuse of synthetic fertilizers and pesticides, but they won't necessarily eliminate either of them. If we want to get rid of these agricultural inputs altogether, we will either need to ban them or (preferably) develop alternatives that farmers will readily adopt.

Although agriculture uses little energy, relative to the rest of our economy, agriculture is a major user of water. Much of the water falls as rain and returns to the atmosphere as evaporation from leaves and soil, but significant amounts are drawn from rivers and underground aquifers. Agriculture accounts for up to 80 percent of this use, by one estimate.[9] In some regions, underground water that accumulated over many decades is being pumped for irrigation faster than it is replaced. This is an example of the second kind of threat to sustainability: depletion of a resource critical to agriculture.

Phosphorus for fertilizer is another example. Worldwide production of this mineral may peak as soon as 2030, while demand continues to increase.[53] Twenty years is not a long time to find and implement solutions to such a complex problem.


## Reliability and Risk

Even if we solve the problem of sustainability and reverse the long-term trends that threaten our food supply, shorter-term variability in food production can be a serious problem. People need to eat every month, not just "on average." A few years ago, my wife and I visited the ruins in Chaco Canyon, New Mexico. These enormous stone buildings, with over 100 rooms each, were abandoned around 1150. A major reason was a 25-year drought. In a museum, we saw evidence for this drought: a series of much thinner annual growth rings in tree trunks from that period.

It is clear that the world is not prepared for a global drought of such duration. Fortunately, droughts usually affect only particular regions. So long as the world overall has significant excess capacity to produce food (especially grain, which can be stored for years if kept dry and protected from pests), food grown elsewhere can be shipped to drought-stricken areas. This assumes that those in the drought-stricken region can pay for the grain, using money saved in good years, or that they receive some kind of international assistance. The less extra capacity we have, however, the greater the risk that bad weather or pest outbreaks in one region will undermine global food security. In 2008, for example, governments of several countries banned rice exports, ensuring their own food supplies at the expense of their usual trading partners.

Catastrophes that decrease food production worldwide are less common than regional disasters, but they are not so rare that we can ignore them completely. In 1816, for example, volcanic dust from the eruption of Mount Tambora in Indonesia caused unusually cold temperatures over much of the globe. That "year without a summer" resulted in widespread crop losses due to frost and many deaths from hunger. We do not know when such a global problem will occur next, but we can be certain that even larger volcanic eruptions will happen occasionally. Yellowstone National Park has a dramatic display comparing past volcanic eruptions, some of which produced hundreds of times more ash than the 1980 eruption of Mount Saint Helens. Major volcanic eruptions can devastate vast areas and may have been responsible for some mass extinctions.[54]

Major asteroid impacts have apparently caused more mass extinctions than volcanoes have. Tracking of objects in space provides earlier warnings than is currently possible for volcanic eruptions, however. We are apparently fairly safe from major extraterrestrial impacts for the next 50 years,[55] giving us time to prepare.

We may not be able to prepare for some events large enough to cause mass extinctions, but can we at least develop food systems that survive more mundane catastrophes? To cope with events that reduce crop production only temporarily, we could simply increase worldwide storage of grain. A 2-year supply would fall short of the 7 years recommended to the Pharaoh by Joseph in the Bible, but it would be a big improvement over the 7-week supply available in 2007.[7]

Maybe agriculture needs to consider only disasters that would decrease food supply more than they decrease human population. It might seem that a global epidemic (perhaps caused by terrorists using the recently resurrected[56] flu virus that killed at least 20 million people in 1918) would alleviate food shortages rather than cause them. But modern agriculture depends on manufactured inputs, global distribution of food and fertilizer, power grids, roads, aqueducts, weather satellites, and law enforcement. We therefore need to consider how epidemics might affect infrastructure. How long might electricity for irrigation pumps be off if key people were sick or staying home caring for sick relatives? War, too, can disrupt infrastructure as well as killing people. To the extent that traditional or organic farming methods are less dependent on fragile infrastructure, maintaining the capacity to return to these methods might enhance food security. I return to this point in chapter 12, when I discuss bet-hedging (see glossary). Of course, anything we can do to make our infrastructure less fragile would also be worthwhile.

In summary, food security is not something we can take for granted. Despite substantial technological progress, from higher-yielding crops to computer-guided tractors, global food production is not keeping pace with global population growth and the increased food demand that comes with growing incomes. Unsustainable practices, resource depletion, and disruption (by bad weather, epidemics, natural disasters, or war) all need to be addressed before we can rely on agriculture to meet human food and feedstock needs, consistently and for the foreseeable future.

## Protecting the Environment . . . and Farmers

We need to consider agriculture's overall impact on the environment, in addition to its ability to supply us with food. Agriculture may have less impact, per acre, than land uses like housing, but agriculture's impact is magnified by its extent, 35 percent of the world's ice-free land surface.[10] To limit negative impacts, we need to limit the land used for agriculture. In particular, we need to limit the conversion of natural ecosystems, such as forests or wetlands, to agricultural use.

Natural ecosystems provide many valuable ecosystem services (see glossary), although I will suggest in chapter 11 that these services may not be the most important reason to preserve nature. One estimate of the economic value of ecosystems services, $12 trillion per year,[57] may include some double-counting. For example, nutrient cycling in estuaries is credited with contributing economic benefits worth $21,000 per acre each year. This might be a reasonable estimate of the cost of technological substitutes for these natural processes, but it is not clear how humans benefit economically from nutrient cycling in estuaries other than through its effects on seafood production or recreation, both of which are credited separately. We don't drink brackish water from estuaries or use it to irrigate crops.

Nevertheless, it is clear that natural ecosystems do provide major benefits to humans. Forests remove carbon dioxide from the atmosphere, thereby reducing global warming with all of its risks (spread of malaria mosquitoes out of the tropics, flooding of coastal cities from melting polar ice, and so on). Both forests and wetlands purify water, benefiting fisheries as well as drinking water.

To what extent could agricultural ecosystems provide the same services as the natural ecosystems they displaced? The answer may depend on what crops and livestock are raised and how. Today, agriculture often makes negative rather than positive contributions to some aspects of environmental quality. For example, nutrient runoff from agriculture (nitrogen mostly from fertilizer use on cropland; phosphorus mostly from animal manure on pastures and rangeland[58]) is thought to be a major cause of the oxygen-free "dead zone" in the Gulf of Mexico. Could agriculture have a net positive impact instead, recycling nutrients from cities that would otherwise end up in the Gulf? Any comprehensive list of goals for agriculture should certainly include enhancing ecosystem services, as well as minimizing pollution and other negative effects on the environment.

Last, we need to recognize the importance of aesthetic, socioeconomic, and cultural aspects of agriculture. Farms can provide habitat for birds and other wildlife, although not necessarily comparable to the forests, grasslands, or wetlands that once occupied the same land. I remember watching large numbers of swallows catching insects over my brother Tom's farm. Many rice farmers in California now manage their fields as waterfowl habitat during the winter, where birds have a less diverse diet but may be safer from predators.[59]

Sometimes, there may be tradeoffs among our goals. In such cases, which goals should have priority? If three bird-rich acres of shade-grown coffee are needed to produce the same amount of coffee as one bird-poor acre of sun-grown coffee plus two acres of rain forest,[60] is the latter option worth considering? If so, how can we keep two different coffee farms from claiming that they are preserving the same two acres of rain forest?

Although this chapter does not focus on farmers' lives, ignoring farmers' needs and preferences would be a mistake. If enough farmers decide that they would prefer a profession with higher pay or better working conditions, that could lead to food shortages. A farmer shortage could be a problem even if all the other problems discussed earlier are solved.

The total number of farmers is not my only concern, however. There are many different kinds of farmers and many different agriculture-related specialties. What sorts of expertise are we most likely to lack when we need it? If shortages of inorganic fertilizers develop, we will need farmers skilled in the use of organic fertilizers or nitrogen-fixing legumes. Labor shortages, on the other hand, could increase the demand for agricultural engineers and farmers familiar with highly mechanized agriculture. Unusually severe crop disease epidemics could create a sudden need for more plant pathologists. Nobody specializes in agricultural problems caused by massive volcanic eruptions, but collaboration among crop physiologists and ecologists, agronomists, soil scientists, and climatologists could speed recovery from such disasters.

The knowledge and skills needed to help solve such problems can take a decade or more of specialized education and real-world experience to develop. A plant pathologist, for example, would typically have taken extra math and science courses in high school, followed by 4 years of undergraduate education (perhaps in plant science), 6 years of graduate education in plant pathology, and 2 years of postdoctoral study before starting her first professional position. Add 4 years of on-the-job experience, for a total of 20 years. So, when crop disease epidemics strike, our ability to respond depends on education funding over at least the previous two decades. Most farmers would have fewer years of formal education than a plant pathologist, but there is a big difference in expertise between a farmer with 5 versus 20 years of experience.

Unless we can be sure which kinds of agricultural expertise we will need 20 years from now, it would be prudent to educate students in a wide variety of agricultural topics. This suggestion is consistent with the bet-hedging theme in chapter 12. A senior scientist with a major seed company recently complained to me that they could hire as many molecular biologists as they needed (presumably because so much research funding has been going to molecular biology), whereas people who can manage field tests are in short supply. We should encourage a wide variety of agricultural professionals, from organic farmers to molecular biologists specializing in the diagnosis of crop disease, to continue updating their skills. Appropriate programs and policies could range from scholarships and research grants to targeted tax cuts or simplification of overly complex regulations.

I am emphasizing diversity of skills rather than numbers. We may not need as many farmers as in the past, because mechanization makes it possible for one person to farm more acres, but we cannot afford to lose expertise we might need in the future. We may not need many weed ecologists, but we do need some. Universities might need incentives to continue offering certain critical classes to a dozen agriculture students per year, when psychology and business classes are enrolling six hundred students.

## Perspective

By the time you read this, some droughts may have ended. Some land abandoned when crop prices were low may be farmed again. Government policies that favor turning corn into fuel (plus, to be fair, protein-rich by-products eaten by animals who might otherwise have eaten that corn) may have been reversed. Any of these could ease food shortages and lower food prices, in the short term.

Agriculture's underlying long-term problems will not be solved easily, however, even if other issues displace agriculture from the front pages of newspapers. Long-term trends of population growth, increased food demand per person (partly from increased consumption of meat or ethanol), yield decreases caused by unsustainable agricultural practices, and gradual resource depletion will all affect agriculture. We need enough spare food-production capacity that natural or human-made disasters are not aggravated by food shortages. We want agricultural landscapes to be safe and pleasant places to work and visit, but we also want some land left as parks or wilderness areas.

A book driven by ideology would attempt to explain how a single solution can simultaneously achieve all of agriculture's goals. But studying evolution helps us see the importance of tradeoffs. A well-known example is the competition, within individual plants, between stem growth (increasing competitiveness for light) and grain production.

Tradeoffs can also occur on larger scales. We may not be able to maximize crop yield, thereby reducing land area needed for crops (so perhaps sparing land for nature, although this is not certain[61-63]) while also minimizing fossil-fuel use per acre. An agricultural system that minimizes food costs in good years, freeing resources for other priorities, may not necessarily be the system that recovers fastest after a disaster. We should certainly try to improve all aspects of agriculture, but sometimes we will have to balance competing goals.

Better farming methods are not a complete solution, but they will make difficult choices easier. We need improved agricultural methods (broadly defined to include new crop varieties and better distribution systems for inputs and food) that are both more efficient and more sustainable. Developing these improved methods will require a sustained increase in agricultural research. In other words, we will need to dedicate a larger fraction of available brainpower and other resources—more people and more money—to solving agricultural problems. These problems are so serious that we cannot afford to neglect any relevant branch of human knowledge, from chemistry to economics. In particular, we cannot afford to ignore insights from evolutionary biology.

# 3

## Evolution 101

### THE POWER OF NATURAL SELECTION

THIS CHAPTER REVIEWS the basic vocabulary and fundamental principles of evolutionary biology. My main focus here is on the power of natural selection to improve the adaptation of individual plants and animals to their environment.

### Evolution by Natural Selection: A Well-Tested Theory

The power of natural selection and the many millennia over which it has operated are key to one of the central themes of this book: *It will often be difficult for biotechnology (genetic engineering) to improve on what past natural selection has already achieved.* There are some important exceptions to this claim, however, that will be explored in later chapters. In particular, we may be willing to accept some tradeoffs that were consistently rejected by past natural selection.

Evolution is a scientific *theory*: a collection of facts and well-tested hypotheses that can be used to explain a wide range of observations and to make accurate predictions. This scientific definition of a theory is very different from the popular definition: a "wild guess." A *hypothesis* is an explanatory statement that makes predictions specific enough that we would reject the hypothesis if those predictions fail.[64] (A short definition of a *hypothesis* is a "potentially falsifiable prediction generator.") The germ theory of infectious disease has largely replaced the once-popular "witchcraft theory," because it provides better explanations of past epidemics and because it correctly predicts how crowding, water treatment, quarantines, vaccination, or antibiotics will affect the spread of disease. The theory of evolution explains why living things are the way they are and can, to some extent, predict how they will change over future generations. Throughout science, correct predictions increase our confidence in a theory, whereas incorrect predictions show that the theory needs to be revised in major or minor ways.

*Peanutbutter Chocolate Ice cream.*

Jerry Coyne's book, *Why Evolution Is True*,[1] is an excellent summary of the evidence that life on earth has evolved and is still evolving. This evidence includes recently discovered fossils, like those linking whales to their hippo-like ancestors; geographic patterns of species distribution consistent with our understanding of how two or more new species can arise from one existing species; and experiments where scientists have observed evolution as it happens. Coyne also discusses genetic comparisons among species, which are an increasingly important source of information about evolution. For example, other primates have the same defective copy of a vitamin-C gene as humans, while guinea pigs have a different defective version. Most other mammals have the functional version. These data are consistent with all primates sharing a common ancestor (with a defective gene) who lived more recently than the common ancestor of all mammals (with the functional gene). If you are interested in the evidence for evolution, particularly how species evolve from other species, Coyne's book would be a good place to start.

My own focus is mostly on changes within species and on their implications for agriculture. The evolutionary changes most relevant to my arguments are mainly the result of the process of natural selection, so this chapter discusses some of natural selection's strengths and limitations. Its most interesting limitations are those imposed by various tradeoffs. This is because tradeoffs often represent opportunities for humans to make improvements missed by past natural selection, as discussed in later chapters.

## How Evolutionary Changes Occur: Definitions and Mechanisms

A species will evolve by natural selection if three conditions are met. First, there must be differences among individual members of the species. Second, individuals must have some tendency to inherit the characteristics of their parents. Third, *inherited differences must affect reproductive success*. The emphasized text is the key to evolution by natural selection.

For a more complete discussion, we need some definitions. A *species* can be defined as a group whose members can interbreed, producing fertile offspring. A species may include many separate populations (see glossary), with much more interbreeding within populations than between them. Genetic inheritance passes mainly via *deoxyribonucleic acid* (DNA), a long molecule (typically coiled up, inside cells) carrying the instructions that guide the growth, development (see glossary), behavior, and reproduction of a plant, animal, or microbe. A DNA molecule is analogous to a string of millions of "beads" (*nucleotides*, symbolized G, T, A, and C) strung together. I will define a *gene* as a section of DNA that codes for a particular function, such as breaking starch down into sugars, or turning other genes on or off. The *genotype* is a collective term for all the genes in an individual. An individual's *phenotype* is the sum of all its observable traits, from height to behavior, which result from the interaction of genotype and environment over time. (The terms *genotype* and *phenotype* were coined by a plant breeder, Wilhelm Johannsen,[65] an example of the central role of agriculture in the development of biological science.) *Alleles* are alternative versions of the same gene, differing in DNA sequence and potentially affecting phenotype.

*Evolution* is simply a change in the relative frequency of alternative alleles in a population or species. For example, an allele that tends to make a tree flower earlier in the spring might become more or less common, perhaps disappearing altogether, relative to another allele that tends to make flowering later. Evolutionary changes may be slow or fast. The effects of two alternative alleles on phenotype may be similar or very different. We refer to the relative success of an allele, its increase in frequency over time, as its *fitness*. Sometimes *fitness* is also used to refer to the relative success of a genotype or an individual, with success defined as an increase in relative representation over generations.

Chance plays important roles in evolution, but evolution is not solely due to chance. For example, processes causing mutations are somewhat random, whereas selection is not. *Mutations* are changes in DNA sequence that happen within individuals and may be inherited by some or all of their descendants. Some mutations may be more likely than others, but mutations are generally random in the sense that beneficial mutations are usually no more likely than harmful ones.

Some mutations have little effect on phenotype. Of those that do have an effect, most are at least slightly harmful, assuming conditions are fairly constant. This is because they represent a change from alleles that were successful in the past. A mutation may be harmful under some conditions and beneficial under others, however. For example, a mutation that changes GAG to GUG at a key point in a particular human gene changes one of the amino acids (see glossary) that are strung together to make the protein hemoglobin. Three successive nucleotides code for one amino acid. This mutation, leading to the sickle-cell trait, decreases susceptibility to malaria, but it also reduces the blood's ability to carry oxygen. Tradeoffs like this are key to the central arguments in this book.

*Selection* (either natural selection or selection imposed by humans) is a change in the relative frequency of alternative alleles in a population, resulting from differences among the alleles in their effects on survival and reproduction in a given environment. Almost all improvements in a species' fit to its environment are due to selection. An *adaptation* is a trait that increases fitness under current conditions and that resulted from past selection.

Selection can be imposed by plant or animal breeders deciding which individuals are allowed to reproduce. Over much of our history, such deliberate selection by humans may have been less common than inadvertent selection caused by humans. For example, seeds harvested by humans and protected over the winter are more likely to grow into seed-producing plants the next year, relative to seeds left on the ground. Thus, early farmers inadvertently selected for plants that hold onto their seeds rather than dropping them, simply by harvesting seed for future planting.

Purely natural selection, with no direct human involvement, was common over most of the evolutionary history of our crop plants. Natural selection benefiting some alleles over others may be imposed by physical conditions, such as drought, or by interactions with other plants and animals.

For example, some clover plants make toxic cyanide when wounded. Plants with an alternative allele do not make cyanide. How does selection change the relative frequency of these two alleles? It depends on conditions. In a warm location, a clover population that started with 9 percent of plants making the cyanide-producing *enzyme* (a protein that speeds a chemical reaction) evolved in one year to have 21 percent cyanide-producers.[66] In a cold location, a similar clover population evolved to have only 3 percent cyanide-producers, also in just one year.

This example illustrates two important points. First, plants can evolve fast enough that we can't assume that this year's weeds are identical to last year's. Species with shorter generation times, including many agricultural pests and pathogens, can evolve even faster. For example, microbiologists Paul Rainey and Mike Travisano, with whom I now collaborate, found that bacteria can diversify from the one ancestral genotype into three distinct types in only a few days.[67]

Second, differences in clover gene frequencies between warm versus cold climates are an example of tradeoffs in adaptation to different environments. Slugs and snails, which are more common in warm locations, impose selection for cyanide production by selectively eating cyanide-free plants. Possible reasons for decreasing cyanide production in cold climates will be discussed in the next chapter.

But natural selection is not always caused by the physical environment, or by other species. Often, natural selection results from competition among individuals of the same species for resources. Both Charles Darwin and Alfred Russel Wallace, who recognized the importance of natural selection at about the same time, attributed this key insight to reading Malthus. Darwin and Wallace realized that every species has the potential for exponential population growth—the same exponential growth that, in humans, concerned Malthus.[32] A growing population will reach the point, sooner or later, where resource supply per individual is not sufficient to support survival or reproduction. For example, if thousands of seedlings were to share the sunlight reaching a square foot of land equally, none of them could photosynthesize enough to produce even one seed before winter. Only seedlings that get more than their share of sunlight, often by growing taller than their neighbors, will survive and reproduce.

Seeds that germinate earlier in the spring get a head start, but they may also be at more risk from frost. In that case, natural selection driven by competition for light opposes natural selection driven by frost, a typical evolutionary tradeoff. The overall direction of evolutionary change due to natural selection will often depend on the relative importance of multiple factors.

*opposing forces* [handwritten margin note]

Natural selection is not the only mechanism of evolutionary change. *Gene flow* is a change in allele frequencies in a local population due to genes arriving from somewhere else. If bees carry pollen from crop plants to wild plants with which they can interbreed (cultivated and wild sunflower, say[68]), then each seed resulting from this cross-pollination will have about 50 percent crop genes. Gene flow will often make recipient populations less adapted to their environment. For example, crop genes for holding onto seeds, rather than scattering them, may reduce the fitness of weeds. This contrasts with natural selection, which improves adaptation. Sometimes, however, gene flow may introduce alleles that prove to be beneficial in a new environment.

*Genetic drift* refers to random evolutionary changes, whereby genotypes that might usually be less fit become more common through chance (such as a seed landing in an unusually favorable spot), or vice versa. Drift is most powerful in small populations, because it is unlikely that hundreds of plants of a given genotype would all get lucky. For example, if a farm field has just been invaded by a small group of weeds (three velvetleaf plants, say), the plant that makes the most seeds may not be the one whose genotype is best suited to that field. It may just be the one plant that the farmer didn't notice and pull up.

*Recombination* mixes genetic material from two parents in new ways. For example, part of a *chromosome* (many genes linked together) inherited from a male parent may be exchanged with the corresponding chromosome from the female parent, resulting in a pair of mixed chromosomes, each with alleles from both parents.

Mutation and recombination affect DNA molecules directly, generating new alleles. Selection, gene flow, and genetic drift affect the frequency of alleles in a population indirectly, by affecting the survival, reproduction, and movement of individuals having those alleles.

## An Example of Evolution by Natural Selection

Figure 3.1 is a simplified example of how populations evolve by natural selection and also illustrates the sort of tradeoffs that are key to my arguments. At the top of the page, a soybean seed arrived on a plateau with no other plants, perhaps in mud on the foot of a duck. In year 1, the seed grows into a plant that produces one pod containing three seeds and then dies. (Annual plants like soybean do usually die soon after reproducing, but real soybeans can make many pods.)

The next row shows year 2. Each of the seeds produced in year 1 has grown into a plant. (This is an unusually high survival rate, reflecting the lack of competition, and somewhat balancing my one-pod assumption.) Two plants are similar to their parent, but one has a new genetic mutation that makes it grow taller and with darker leaves. One plant has this particular mutation, or 33 percent of the tiny, three-plant population. The cost of making a taller stem leaves the plant with fewer resources to make seeds, so it makes only two seeds.

In year 3, every seed again grows into an adult plant. The tall/dark mutation affects reproductive success in ways that may depend on conditions. So far, tall mutants have reproduced less than the shorter plants, so their frequency decreased from 33 percent in year 2 to 25 percent of the population in year 3. This change in frequency means that the population has evolved.

By year 3, it's starting to get a little crowded on the plateau. The plants at the edge of the plateau still get plenty of light from the side—this is an example of an *edge effect*—but plants in the middle are now shaded by their neighbors each morning and afternoon, when the sun is low in the sky. Photosynthesis depends on light, so only plants at the edge still make three seeds. Short plants unlucky enough to have the tall mutant as a neighbor get shaded more, so they make only one seed or, with two tall neighbors, none at all. With increased crowding, average seed production by shorter plants has become lower, rather than higher. As a result of this reversal in the direction of natural selection, the frequency of taller plants increased slightly, to 28 percent in year 4.

In year 4, crowding has become severe, and I have stopped drawing the family tree connecting parent plants with their offspring. The plant population has reached *carrying capacity*, the maximum value the plateau can support. The 14 plants produce 14 seeds, so population is no longer increasing. But the genetic composition of the population is still changing, because tall plants shade their shorter neighbors, reducing their seed production.

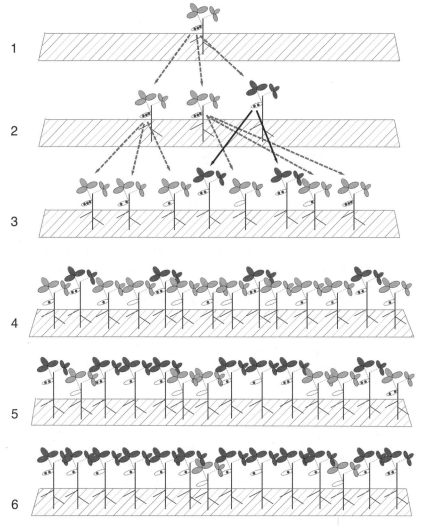

Figure 3.1. An example of evolution by natural selection, showing how changing conditions (for example, crowding) can change the direction of selection.

In year 5, 8 of 14 plants (57 percent) are tall. Only one short plant produces any seeds. But now there are enough tall plants that they are starting to shade each other. Although a tall plant competing only with short plants (year 3) could make two seeds, by year 5 tall plants are paying the cost of making those tall stems, without getting any more light than if they were all short. With the same photosynthesis as shorter plants—being taller doesn't get you more light, if your neighbors are also tall—but a higher cost for making tall stems, many of the tall plants now make only one seed.

In year 6, only two short plants are left, so 86 percent of the population is tall. Both short plants die without reproducing, so the population will be 100 percent tall in year 7.

Here are some key points from this example. The size of the soybean population on the plateau increased and then stabilized at the carrying capacity of the plateau. At the same time, the genetic composition of the population (tall versus short) also changed. Changes in population size and genetic composition, respectively, are the domains of the sister disciplines of ecology and evolution.

Population density and population composition interacted. At low population density, the cost of making stems made tall plants produce fewer seeds than shorter plants, so the frequency of tall plants decreased. As the plateau became more crowded, however, competition for light intensified. Because taller plants got more than their share of light, they photosynthesized more than shorter plants. Shorter plants shaded by their taller neighbors made fewer seeds, so they gradually disappeared from the population. As tall plants increased in frequency, however, they started competing mainly with each other. So they no longer got more light per plant. Because they still paid the higher cost of taller stems, they actually produced fewer seeds per plant than the shorter plants they displaced.

This is an example of a tradeoff with enormous agricultural implications: sometimes the genotypes favored by natural selection (for example, taller plants) are different from those that, collectively, produce seeds more efficiently (shorter plants). In such cases, plant breeders can increase the efficiency of crops by reversing some of the effects of past natural selection, as discussed in chapter 8.

### Complex Adaptations Evolve, with No Long-term Goal, in a Series of Steps

I will argue in chapter 5 that many of the genetic changes proposed by biotechnologists would produce phenotypes that have already arisen repeatedly, through mutation, but been rejected by natural selection. This argument hinges on an adequate frequency of mutations and on the power of natural selection.

It is true that any specific single-nucleotide mutation (a particular G changing to a T, say) will arise infrequently, perhaps once or twice in a hundred million individuals.[69,70] For common species, however, that low frequency may not slow evolution much. How much land would we need to have a hundred million plants? Farmers often grow a hundred thousand soybean plants per acre, so a hundred million plants would occupy only a thousand acres, or about 4 square kilometers. Therefore, for any abundant plant species with widespread distribution, any given single-nucleotide mutation is likely to arise many times per year. Also, there will often be many different mutations that would result in the same phenotypic improvement.

Natural selection isn't limited to improvements that can be achieved through one single-nucleotide mutation, but it certainly hasn't tested every possible genotype. Evolutionary changes via natural selection depend on current conditions (that is, on which of the existing genotypes are most fit right now) rather than some distant goal. An analogy with a river may be useful. Water flowing down from the mountains may eventually end up in an ocean, but it is not striving toward the ocean as a goal. A river may flow in various directions, including away from the ocean, as it responds to the local slope of the land. And there's no guarantee that every river will reach the ocean. A river may flow into a basin with no exit and form an inland lake.

Similarly, evolutionary changes in allele frequencies in populations respond to current conditions, whatever the long-term consequences. A population evolving by natural selection alone (that is, ignoring mutation and drift) will never abandon genotypes with currently higher fitness for genotypes with currently lower fitness, even if the lower-fitness genotypes could be seen as stepping stones to genotypes with much higher fitness.

How much does the requirement for increasing fitness in each generation limit the potential of natural selection? There is some disagreement about this,[71] but there have been few detailed molecular studies of multistep evolution so far. Recent advances in gene-sequencing technology may greatly accelerate such research. Meanwhile, consider a study by evolutionary biologist Daniel Weinreich and colleagues. They studied bacteria that had evolved resistance to a particular antibiotic.[72,73] They analyzed the DNA sequences of resistant and susceptible genotypes and found that their DNA differed in five separate places.

In a population of millions of bacteria, there is a good chance of finding mutants with each of these differences separately, but all of them at once? Very unlikely. So they assumed that the change from susceptible to resistant took place in a series of five steps. If the fitness of intermediate genotypes didn't matter—that is, if populations could sacrifice fitness for a few generations in pursuit of distant goals—then the five steps could have occurred in any order—that is, by any one of $5 \times 4 \times 3 \times 2 = 120$ possible pathways. But for the population to have evolved by natural selection alone, none of the steps could have involved a decrease in fitness.

Given that antibiotic resistance did in fact evolve, they reasoned that at least one pathway without lower-fitness intermediates must exist. They set out to find all possible pathways that could have been followed by natural selection. There are 32 genotypes to consider ($2^5$ possible combinations of five mutations), so they made bacteria with each of the 32 genotypes, using genetic engineering. Then they measured the fitness of each genotype in the presence of the antibiotic. Based on those measured fitness values, they found that 102 of the 120 possible pathways required at least one step from greater to lesser fitness; those pathways are inaccessible to natural selection. But there were 18 different pathways that never passed through intermediates with lower fitness. Any of those pathways provided a plausible route for antibiotic resistance to evolve, with each of five successive steps favored by (or at least allowed by) natural selection. This is consistent with our understanding of how natural selection improves adaptation: an evolutionary change that would be highly unlikely in one step can occur in a series of small steps.

This bacterial example shows that moderately complex adaptations can occur in a series of steps, with each step favored by natural selection. But how do more-complex adaptations, like clover leaves that turn to track the sun across the sky, evolve? Was there really an evolutionary pathway from nontracking to tracking plants that increased fitness at each step?

There are at least four ways that adaptations can evolve via intermediates that are usually less fit. First, the relative fitness of different genotypes can fluctuate from year to year, as different conditions favor different genotypes. Therefore, even evolutionary paths involving intermediates that are usually less fit may not always be closed to natural selection. In Weinreich's antibiotic-resistance experiment, for example, it is conceivable that some of the evolutionary paths that were blocked by lower-fitness genotypes, under their experimental conditions, would actually have been higher-fitness pathways under other conditions, such as a different temperature.

Second, genetic drift can change the genetic composition of small populations (a few isolated plants, say) in ways that natural selection alone cannot. A mutant genotype might tend to disappear in competition with its ancestral genotype. But what if one or two individuals with the inferior mutant genotype happened to colonize an area where there was less competition? Then their descendants might survive for enough generations that additional mutations and natural selection, together, would yield a genotype more fit than any in the source population.

Third, animals and plants that reproduce sexually inherit two (possibly different) copies of most alleles, one from each parent. An individual with one mutant allele that is defective for its original function won't necessarily have lower fitness, if the other allele is normal. So "defective" alleles may survive for enough generations to mutate further, perhaps gaining some useful new function.

Fourth, evolution may sometimes proceed via intermediates that have consistently lower fitness, even though natural selection consistently reduces their frequency during the intermediate generations.

## Evolution via Less-fit Intermediates

Fyodor Kondrashov and colleagues have recently found an unusually clear example of two-step evolution in mammals, where less-fit genotypes survived long enough to have their fitness restored by a second mutation.[74]

They studied mutations in a gene found in *mitochondria*, energy-producing organelles inside animal and plant cells, which retain some of the genes inherited from their distant ancestors, symbiotic bacteria.[75,76] Because mitochondria are inherited only from mothers, there's no spare copy from the father to mask the fitness effects of a defective copy. The particular mutations they studied affected the stability of transfer RNA (tRNA; see glossary) in a way whose fitness consequences are particularly easy to understand.

Transfer RNA molecules have a key role in protein synthesis. Each tRNA binds to a different three-base sequence in a messenger RNA, via the complementary triplet at one end of the tRNA (for example, GUA on the bottom loops in figure 3.2A). The other end of the tRNA carries the corresponding amino acid, which is tyrosine for the tRNA shown. Transfer RNA thereby translates the sequence of bases in messenger RNA (transcribed from DNA) into the amino acid sequence in proteins.

Transfer RNA molecules are held together by bonds similar to those that hold the two strands of the DNA double helix together. U (the RNA equivalent of T) bonds with A, while C bonds with G. If two simultaneous mutations changed a UA bond to a CG bond—compare the left and right sides of figure 3.2A—then the double-mutant transfer RNA would still be stable. But simultaneous mutations are rare. So a lineage that acquires only one such mutation (U → C, say) would have to limp along with lower-stability tRNA, perhaps for several generations, until it acquired the compensatory mutation in the mutant nucleotide's partner (A => G), restoring stability.

We might expect such lower-fitness lineages to go extinct before re-evolving stability, but Kondrashov and colleagues found evidence for many cases of successful two-step evolution of tRNAs. They used *ancestral-state reconstruction*,[77] a set of sophisticated mathematical tech-

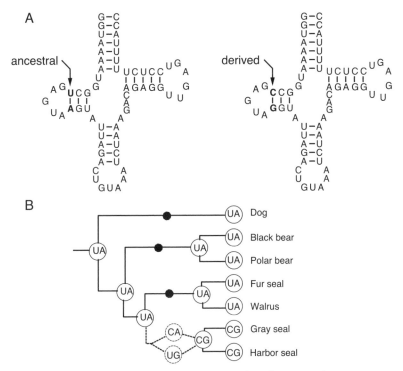

Figure 3.2. Evolution of transfer RNA genes via less-fit intermediates. Based on information from Fyodor Kondrashov.

niques for determining the genotypes of extinct ancestors from their modern descendants.

They began with evolutionary family trees, such as the phylogeny (see glossary) for dogs, bears, seals, and walruses shown in figure 3.2B. The tRNA sequences for these living species were determined directly, but what about their extinct ancestors? In particular, which type of bond did the last common ancestor (see glossary) of these species have at the position indicated by the arrows in figure 3.2A: UA or CG? (Parents are the last common ancestors of siblings; grandparents are the last common ancestors of first cousins.)

The simplest hypothesis consistent with the data for living species is that their last common ancestor (represented by the circle on the left in figure 3.2B) had genotype UA. That hypothesis requires only a single UA → CG transition (via CA or UG), before the last common ancestor of gray seals and harbor seals, but after the last common ancestor that these seals share with fur seals and walruses. An alternative hypothesis, that the last common ancestor of all seven species had genotype CG, is less likely. This is because that hypothesis would have required at least three CG → UA transitions, at the points indicated by the solid circles. A more complete version of the phylogeny in figure 3.2B is included in the paper by Kondrashov and colleagues and further supports UA as the ancestral state.[74]

Figure 3.2B shows only one two-step evolutionary transition, but by analyzing data for 83 mammal species, with many different tRNAs (corresponding to different DNA triplets and amino acids) for each species, the authors found over one thousand examples of such transitions. Each of these must have passed through a lower-fitness intermediate, as shown in the ancestry of gray and harbor seals in figure 3.2B. So lower-fitness intermediates are not an absolute barrier to evolutionary improvements, although they do make any given improvement less likely.

## Natural Selection Has Had Plenty of Time to Work

So natural selection isn't limited to single-step improvements. Complex multistep improvements can easily proceed via natural selection, so long as each step is an improvement, as in the five-step evolution of antibiotic resistance. Although evolution has no long-term goal, lower-fitness intermediates are not an insurmountable barrier, as seen with transfer RNA. Natural selection is quite good at finding genotypes with greater fitness, if the path from current genotypes isn't too complicated. But has evolution been going on long enough to produce complex adaptations, requiring a long series of steps?

For example, consider some ancestral plant that always grew its taproot to about the same depth, whether conditions were wet or dry. An alternative genotype that adjusted its rooting depth, growing shallower in wet years and deeper in dry years, might have greater fitness than any genotype with fixed rooting depth. But what if evolving that superior trait would require significant changes to several different genes? This could require a series of steps, possibly taking thousands of years. Has evolution had long enough to make such sophisticated improvements?

There are really two aspects to this question. First, has there been life on earth long enough for major evolutionary improvements to have occurred, mostly by natural selection? Second, have conditions remained similar enough over time to keep natural selection working in a consistent direction?

The earth is about 4.5 billion years old, as shown by isotope analyses.[78] These methods have been checked in various ways. One clever approach used daily and yearly growth rings in corals. Modern corals have about

these things evolved w/ along w/ Abiotic factors + Soil other Plts + Animals herbivores ...etc.

Its more than just mutations also its environmental pressures

365 daily growth rings per yearly growth ring, consistent with our current 24-hour day. But fossil corals that died about 400 million years ago (based on isotopes) have more daily growth rings per year. This difference shows that, when the fossil coral was alive, the earth was rotating on its axis faster. A day lasted only 22 hours, so there were more days per year.[79] Astronomers have measured the rate at which the earth's rotation is slowing, leading to an estimate that 22-hour days occurred about 400 million years ago, consistent with the estimate from isotopes.

Four hundred million years is plenty of time for any of the evolutionary innovations discussed in this book to evolve. For example, it has been estimated (with "pessimistic" assumptions) that a species with a light-sensitive spot could evolve into a species with sophisticated eyes like those in humans and octopi in only a few hundred thousand years.[80] Their calculations assumed that individuals consistently benefit from obtaining information about their environment, such as whether something is moving nearby. Eyes do, in fact, appear to have evolved repeatedly. Each of your eyes has a blind spot where optic nerves pass through the retina, on their way to connecting to the front side of the light-detecting cells. Octopi don't have blind spots, because their optic nerves connect to the back side of those cells.

The consistency of natural selection over time is a harder question. Insect pests of crops may have been exposed to some new synthetic pesticide for only a few years, but they have been exposed to the natural pesticides plants make to defend themselves for much longer. Plants have had chemical defenses of some sort for millions of years, but those defenses can change significantly in less than 100 years of evolution.[81] So any insect adaptations that work against a wide range of toxins have been strengthened by natural selection for millions of years, whereas adaptations to some specific toxins may have a much shorter evolutionary history.

Similarly, plants may have had only a brief exposure to any increased droughts caused by recent climate change, but their ancestors have probably experienced at least occasional water shortages for many millennia. Even wet areas sometimes have dry years, so most of today's plants are descended from ancestors that sometimes had to compete for limited water supplies.

How long has natural selection been improving trait X? Many millennia, if X increases fitness in a wide range of environments. Not so long, perhaps, if X is beneficial only under recent conditions. So traits that are beneficial under a wider range of conditions have longer histories of improvement by natural selection. This will be a key point when we try to decide whether to keep or copy some natural adaptation, versus trying to improve on nature.

## Perspective

Even over millions of years, natural selection hasn't necessarily found the best possible solutions to the various challenges faced by our crops or their wild ancestors. If the ideal structure for an enzyme molecule, a leaf, or a tree is radically different from anything that has actually existed, then natural selection has probably never had the opportunity to increase its frequency, relative to less-ideal alternatives. So I don't assume that natural selection has led to perfection, by any standard. But even less-than-perfect ideas from nature may be useful, if they're better than the ideas we would have without studying nature.

Even "best possible solutions" may disappoint us, however. Various tradeoffs make it difficult or impossible to improve some crop or livestock traits, without at least some deterioration in other traits, as discussed in the next chapter.

# 4

## Darwinian Agriculture's Three Core Principles

WITH AGRICULTURE'S GOALS and the power and limitations of natural selection as background, here are three proposed core principles for Darwinian agriculture. A unified field-crop theory would require additional principles from ecology, soil science, and other fields, but these principles and their implications will be the main focus of the rest of this book.

1. Prolonged natural selection rarely misses simple, tradeoff-free improvements.

    This principle hinges on my definitions of *simple* and *tradeoff-free*, as explained in the following section, and leads to several corollaries. For example, when natural selection has already balanced tradeoffs in ways consistent with our own goals, keeping or copying natural selection's innovations is usually our best option. On the other hand, accepting certain tradeoffs rejected by natural selection can sometimes allow rapid progress. Radical innovations never tested by natural selection can also be worth exploring.

2. Competitive testing is more rigorous than testing merely by persistence.

    Some landscape-scale patterns in natural ecosystems have demonstrated their sustainability by persisting for millennia. However, only individual adaptations have been tested by competition among plants or animals with alternative adaptations. For example, the timing of flowering in wild rice today reflects success in past competition among plants, but the prevalence of near-monocultures of wild rice in natural lakes[154] is not the result of competition among lakes differing in aquatic plant diversity. So ideas inspired by landscape-level patterns of even ancient ecosystems will require more testing, relative to ideas inspired by the competitively tested individual adaptations of wild species. Understanding individual adaptations of wild species may, however, require studying them in the natural ecosystems where they evolved.

3. We should hedge our bets with a greater variety of crops— and ideas!

*yay
OSU
land grant*

> This bet-hedging will require allocating some land and other resources to crops and research programs that seem less promising today but that may outperform today's winners if conditions change.

The first two principles build on a 2003 paper with my PhD student Toby Kiers and visiting evolutionary biologist Stuart West, now at the Free University of Amsterdam and Oxford, respectively.[28] Both principles are intended as hypotheses, subject to possible disproof by counterexamples.[64] The third principle could be considered an opinion rather than a hypothesis, because it may not make specific enough predictions to be falsifiable. If the first two hypotheses are correct, then current efforts to improve agriculture that ignore their implications are risking failure. On the other hand, basing 100 percent of our efforts on those two hypotheses would also violate the bet-hedging principle.

In addition to these proposed principles, agriculture, Darwinian or otherwise, inherits various principles from other disciplines. These include conservation of matter for each chemical element, from physics and chemistry, and the tradeoff principle—"having more of one good thing usually means having less of another"—from economics.[82]

## Prolonged Natural Selection Rarely Misses Simple, Tradeoff-free Improvements

When biotechnologists promise to improve crop traits like drought tolerance, they rarely discuss tradeoffs. But I hypothesize that *prolonged natural selection is unlikely to have missed simple, tradeoff-free improvements*. If this is true, then many of the simple "improvements" proposed by biotechnologists have already been rejected repeatedly by past natural selection, due to various tradeoffs. If we ignore such tradeoffs, we risk making things worse rather than better.

Claiming that natural selection rarely misses simple, tradeoff-free improvements is far from claiming that natural selection has produced perfection. The validity of this hypothesis hinges on my definitions of *simple* and *tradeoff-free*. By a *simple* change, I mean any change that has arisen repeatedly over the course of evolution (typically by mutation), whether or not it was retained by natural selection. By *tradeoff-free*, I mean "tending to increase fitness under all conditions." With these definitions, the hypothesis is almost a tautology. But determining whether a particular change is simple and what tradeoffs it involves is not always easy.

Simple changes include, but are not limited to, any change achievable by a single-nucleotide mutation. Often, there will be many different single-nucleotide mutations that give similar phenotypes. For example, increased expression of a particular gene could typically come from any of several single-nucleotide changes, either in the promoter region for the focal gene (where the enzyme RNA polymerase binds before copying the DNA to messenger RNA), or in other genes that code for transcription factor proteins that regulate this copying process. Any reasonably large population, like the hundred million soybean plants in 4 square kilometers, is likely to have at least one of each possible single-nucleotide mutation,[69,70] yielding several plants with the mutant phenotype in every 10 square kilometers.

Changes qualifying as simple, by my definition, aren't necessarily those that are easiest for biotechnologists. It may be easier for them to add an entire gene from an unrelated species than it is to modify a specific DNA nucleotide in an existing gene. If adding a novel gene results in a phenotype never previously tested by natural selection, then the resulting change isn't simple, and we can't assume that this option has been rejected repeatedly by past natural selection. On the other hand, changes that require a series of steps can readily occur by natural selection, so long as each step increases fitness.

Genetic changes that don't qualify as simple occur less frequently, by definition. Some phenotypes would require so much genetic change that they have never arisen. Or they may have arisen occasionally, but not often enough for the resulting phenotypes to be tested adequately by natural selection. For example, some hypothetical, once-in-evolution event (perhaps involving nearly simultaneous changes to two or more genes) might give an algal cell twice the photosynthesis rate of its competitors. But that cell could get eaten before it has time to reproduce. If a change hasn't been *repeatedly* tested by natural selection, then it may be difficult for us to predict whether it would be an improvement.

By *tradeoff-free improvement*, I mean a change that improves fitness under all conditions. For example, a hypothetical tradeoff-free improvement in cold tolerance would increase survival and reproduction in cold weather, without reducing survival and reproduction in hot weather.

Early in the history of life, there may have been many opportunities for simple, tradeoff-free improvements, such as increasing drought tolerance without decreasing flooding tolerance. But as each "low-hanging fruit" was picked (that is, as each improved allele replaced older alternatives), options for *further* improvements that were both simple and tradeoff-free became increasingly rare. Our first core principle implies that, today, the remaining simple opportunities for further improvements involve accepting tradeoffs rejected by past natural selection.

When might accepting tradeoffs rejected by past natural selection make sense? Consider tassels, the pollen-producing male flowers at the tops of corn plants. Tassels consume resources that could have been used for seed production. They also cast shadows that reduce photosynthesis by leaves below.[83]

Would smaller tassels be better? Much more corn pollen is produced than is needed to fertilize female flowers and produce seeds. In forests, too, Darwin referred to the "astonishing waste of pollen by our fir-trees."[26] But there must be reasons why natural selection has maintained such large tassels.

Large tassels, peacock tails, and the huge antlers of some male animals have all been maintained by natural selection for similar reasons. Plants that produce more pollen fertilize more of the female flowers on neighboring plants, so they have more descendants. And having more descendants is what natural selection is all about. It's also worth noting that the shadows cast by tassels often fall on neighboring plants, so an individual plant doesn't pay the full photosynthetic cost of its own tassel.

Plant breeders have reduced tassel size more than 50 percent since 1930. This was apparently a side-effect of selection for yield, rather than "direct selection."[84] In this case, a tradeoff that prevented natural selection from reducing tassel size didn't matter to plant breeders. We don't care which individual plants pollinate most, so long as there is enough total pollen for good seed production.

We can see tradeoffs either as constraints or as opportunities. To the extent that our current goals are aligned with past natural selection, our ability to improve traits like crop drought tolerance through simple genetic changes (as defined earlier) will face the same constraints that limited improvements of those traits by natural selection. But when past natural selection has operated at cross-purposes with our current goals, it may be relatively easy to improve crops in ways beyond the reach of natural selection.[28]

First, let's consider some cases where tradeoffs make it difficult to improve on natural selection. Then I will give a brief overview of cases where we can improve today's agriculture by accepting tradeoffs that were rejected by past natural selection.

*Tradeoffs as Constraints*

Plant breeders and biotechnologists are still constrained by the same law of conservation of matter that has always constrained improvement by natural selection. Conservation of matter applies to individual elements, like carbon or nitrogen, not just total mass. A given carbon atom can be used to make cellulose to strengthen a stem, nectar to attract pollinators, toxins to repel harmful insects, or starch in seeds to give the next generation an advantage in early competition with neighboring seedlings. But a plant can't use the same carbon atom for all of these purposes at the same time. For example, all else being equal, breeding for larger seeds is breeding for fewer seeds.

Similarly, a given nitrogen atom can't be in two places at once. Nitrogen is an essential ingredient in proteins. So, if nitrogen supply is a constant, neither natural selection nor biotechnology can increase the protein concentration in seeds without decreasing seed size or number. Sometimes, all else may *not* be equal, however. For example, if we can increase a plant's nitrogen supply (via fertilizer, manure, nitrogen-fixing green manures, or a better root system), we may be able to increase seed yield and seed percent protein simultaneously. So far, however, breeding for increased corn yield has been accompanied by inadvertent decreases in grain protein concentration.[84]

Some tradeoffs apparently linked to conservation of matter actually have more complex explanations. Cyanide production by plants may be an example. As discussed in the preceding chapter, natural selection increases the frequency of alleles for cyanide production in clover populations in warm weather, apparently because cyanide protects plants from slugs and snails. On the other hand, cyanide production evolved to be less common in a colder climate. What tradeoffs reduce the frequency of cyanide-related alleles in cold climates?

Every chemical a plant makes consumes resources, such as carbon or nitrogen. Making cyanide a plant doesn't need therefore reduces resources available for other purposes, such as seed production. However, there is apparently more to clover's reduced-cyanide story than resource use alone. Cyanide-producing clover plants make fewer flowers, qualitatively consistent with what we would expect from the resource costs of cyanide and of flowers. But the resource cost of making cyanide can be calculated, as can the resource cost of making flowers. Decreases in flower production linked to cyanide production are much greater than can be explained by the increased resource costs of cyanide production.[85] Therefore, cyanide production must have other costs, besides just resources consumed.

There are many ways that toxin production can be costly to plants, above and beyond resource costs.[86] For example, toxins may repel pollinators as well as pests. Insect pests that are resistant to a particular toxin may seek out plants that contain it, where there will be less competition from other insects. As for cyanide, it may poison the clover plants themselves a little, as well as poisoning the slugs.

Given the various costs of toxin production, it is usually better to produce toxins only if a plant is actually under attack. This is, in fact, what many plants do. Clover plants carry out the early steps in cyanide synthesis, but they don't actually produce the cyanide itself until they are wounded by a slug or snail. Other plants start making defensive chemicals as soon as one of their neighbors is wounded, apparently responding to gases released by injured plants nearby.[87]

Some biotechnologists have suggested that we can increase pest resistance by turning existing inducible defenses (see glossary) on all the time, making them constitutive (see glossary). Maybe, but we should first recognize that this option has probably been rejected, repeatedly, by past natural selection. Evolving always-on defenses from inducible defenses is presumably a simple change, achievable through various mutations that knock out some part of the complex mechanisms that regulate this response. So always-on cyanide production already would have evolved, if it were consistently beneficial. It is possible, of course, that always-on defenses are the best option today, even though inducible defenses were best in the past. This could be true, for example, if pests are almost always present in some agricultural setting.

Here's another example of an inducible defense that involves tradeoffs. Many oat varieties resistant to crown rust are highly susceptible to Victoria blight.[88] This is particularly true of varieties whose crown rust resistance is based on the defensive *hypersensitive reaction*, whereby oat cells that detect the crown rust fungus essentially commit suicide. Their suicide blocks the spread of the fungus, which can flourish only in living host-plant cells.

Why does this defense mechanism, which works so well against crown rust, make oats *more* susceptible to the Victoria blight fungus? That fungus consumes *dead* plant cells. So it makes a molecule that resembles the one the oats use to detect the crown rust fungus. The resulting cellular suicide creates ideal conditions for the Victoria blight fungus.

Is there any way around this tradeoff between crown rust resistance and Victoria blight sensitivity? Maybe we could develop oat plants that use a different molecule—one not made by the Victoria blight fungus—to detect the crown rust fungus. But tradeoffs between resistance to fungi that require living cells and susceptibility to those that consume dead cells seem to be common.[88] Is there a general solution, not specific to a particular fungus?

It turns out that most fungi that attack leaves require living cells, while fungi that attack roots mostly consume dead cells. So maybe biotechnologists could figure out how to express hypersensitive reaction genes only in leaves. However, natural selection has already done that, at least in some species.[89]

## Tradeoffs as Opportunities

What about tradeoffs as opportunities? In chapters 8 and 9, I will discuss opportunities to improve cooperation among plants or among species, based on certain tradeoffs rejected by past natural selection. But let's start with a simpler example: tradeoffs between past and present conditions. Suppose a plant population whose ancestors were exposed mainly to crown rust finds itself exposed mainly to Victoria blight? This could happen either if the Victoria blight fungus invades a new area or if a crop is grown in a new area. Either way, letting natural selection operate on the crop would eventually lead to improved Victoria blight resistance, sacrificing resistance against the locally rare crown rust fungus. Plant breeders can often speed this process, perhaps by crossing local varieties with a variety known to be resistant.

Similarly, today's crops may be exposed to different temperatures than their ancestors were, either due to climate change or because they are now grown in new areas. Again, natural selection would eventually improve adaptation to local conditions, but plant breeding can speed the process. This might involve a change in flowering date, for example.

If we breed crops for tolerance to drought or cold, should we expect tradeoffs? Perhaps not always. If the crop's ancestors never experienced drought or cold, then some drought- or cold-tolerant genotype may not have evolved because it was never beneficial, not because its evolution was constrained by tradeoffs. Usually, however, current levels of drought susceptibility in plants shaped by past natural selection (including the wild ancestors of today's crops) will reflect some balance among various tradeoffs. We therefore need to identify those tradeoffs and decide whether we are willing to accept them.

Sometimes, past natural selection was based on conditions that no longer exist. For example, evolution may not be keeping up with increasing atmospheric carbon dioxide concentrations. Today's plants are apparently adapted to much lower, preindustrial $CO_2$ levels.[27] Tradeoffs linked to performance under past $CO_2$ levels may be limiting photosynthesis today. A plant breeding or biotechnology program that recognizes this tradeoff might be able to increase photosynthesis under current conditions.

Humans, too, may be better adapted to past conditions than to present ones. Modifying our own genomes seems risky, but there are other options. In particular, we can use information about tradeoffs and past environments to help us design healthier lifestyles. For example, tradeoffs between early reproduction and longevity are predicted by evolutionary theory,[90, 91] and have been documented in many species, including fellow primates.[92] Natural selection can favor phenotypic plasticity (see glossary), responding to environmental cues in ways that favor longevity over reproduction, or vice versa, depending on conditions. Certain "famine foods" (such as plants containing natural defensive toxins) may trigger physiological responses that favor greater longevity over early reproduction. This is because whenever ancestral populations were shrinking during famines, children produced later made a larger proportional contribution to the smaller gene pool (see glossary).[29, 93] Would we be willing to sacrifice teen pregnancy to delay aging?

## Competitive Testing Is More Rigorous Than Testing Merely by Persistence

Now, let's turn to our second core principle. In seeking nature's wisdom, we are looking for ideas that have been tested over time. But different aspects of nature have been tested over time in two different ways, either through competition, or just through persistence.

Entities (species, businesses, ideas, and so on) that have survived repeated competition will usually have superior properties, relative to those that merely survived, without having to compete. For example, conch shells are more fracture-resistant than rocks are, because conchs have competed for millennia to be less easily eaten by predators than their neighbors are. I will argue in chapter 6 that natural ecosystems (forests, say) don't compete against each other the way individual plants and animals do. Like rocks, forests have been tested only by endurance. Forests thereby differ from trees, which have been tested more rigorously, through repeated competition. We may find good ideas either in the adaptations of wild species or in the landscape-scale patterns we see in forests and other natural ecosystems. But the latter will require more testing, because the former have already been tested repeatedly through competition.

In seeking nature's wisdom, should we focus on the forest or the trees? Consider natural forests in Iran. Some are mixtures of wild almonds and wild pistachios, while others are almost pure stands of pistachio.[94] Both types of forest have persisted for thousands of years. Therefore, if we assume that ancient ecosystems have been perfected over time, both almond-pistachio forests and pistachio-only forests must be perfect. So which kind of forest should agriculture copy?

Both types of forest have demonstrated their sustainability, simply by persisting for so long. But do we have any reason to think that they would be *less* sustainable if they had some other ratio of pistachios to almonds? Or if we added some walnuts? Or if tree species were arranged in a different pattern? Trees may be clustered by species because nuts fall close to the tree, not because forests with clustering won a competition against forests without clustering.

Also, while sustainability is essential, what about other measures of ecosystem performance? What about wildlife populations supported by a forest, the quantity of nuts that can be harvested sustainably each year, or the quantity and quality of water draining to nearby rivers? Have all of these have been optimized by some natural process? I will argue that we can't *assume* that any particular forest is perfectly organized to deliver any of these benefits. We could, however, compare various forests and try to find relationships between their organization (such as species diversity or spatial patterns) and their performance.

Meanwhile, let's ask similar questions about the individual trees. Both almonds and pistachios have *alternate bearing*. That is, they produce heavy nut crops only in alternate years. Apples have some tendency to do this too, a tendency that could presumably be increased through plant breeding. Would that be a good idea?

Unlike whole forests, individual trees have competed against each other, over many generations. Sometime in the distant past, alternate-bearing mutants competed against otherwise similar trees with annual bearing . . . and won. So at least by the criterion of individual fitness under past conditions, alternate bearing was superior to annual bearing.

That doesn't necessarily mean that alternate bearing will meet our needs today better than annual bearing would. But if alternate bearing was the winner in past competitions, then it must have some useful function or functions. We can therefore proceed to identify those functions— reducing losses to seed-eaters may be one[95]—and determine whether those functions contribute to our agricultural goals today.

## We Should Hedge Our Bets with a Greater Variety of Crops—and Ideas!

I don't expect this third principle to be as controversial as the first two principles, so I won't devote as much space to it. My main point is that bet-hedging involves tradeoffs, but we should often accept those tradeoffs.

World food security depends mostly on three crop species: corn, wheat, and rice. Each species is represented by many varieties, each resistant to different pests and diseases. So it's unlikely that a worldwide crop-disease epidemic would destroy any one of these crops. Still, I worry that we are allocating so little land to alternative crops—even well-known species, like potato, lentil, or buckwheat—that they would be unable to fill the gap if one of the major crops failed.

Similarly, we are betting our future on the assumption that biotechnology will be able to deliver on its promises. We are allocating so few research resources (money and brainpower) to alternative approaches that they will be unable to fill the gap if biotechnology falls short.

If the risks of failure for alternative crops and alternative approaches are no worse than the risks for major crops and biotechnology, then diversifying our crops and research portfolio is obviously a good idea. It's really only bet-hedging if we allocate some resources to options that seem somewhat less promising. Why would we do that? Because while those alternatives may be more likely to fail, individually, than our number-one choices, they aren't all likely to fail *at the same time.*

## Perspective

If some phenotype has arisen repeatedly through mutation, yet remained rare, then individuals with that phenotype must not have greater fitness, at least not consistently. In other words, natural selection is unlikely to have missed simple, tradeoff-free improvements. This implies that some of the "improvements" proposed by biotechnologists are likely to come with significant tradeoffs, as discussed in the next chapter.

Various constraints, including those linked to conservation of matter, result in tradeoffs in adaptation to different conditions. Tradeoffs between individual fitness and our agricultural goals are also important. When natural selection works (or has worked) against our goals, we may be able to make improvements missed by natural selection.

Improvement by natural selection depends on competition. Individual plants and animals have competed against each other, directly or indirectly, and more-fit genotypes have displaced less-fit ones. Therefore, the adaptations our crops and livestock have inherited from their wild and domesticated ancestors are not just "good enough"—they are better than the many alternatives against which they have competed. So it's probably safe to assume that, despite advances in biotechnology, it will be difficult to improve on many of these evolution-tested adaptations.

Natural ecosystems haven't competed against each other the way individual plants and animals have. So we will probably find that natural selection has left some opportunities to improve the organization of landscapes and interactions among species. For example, although cooperation among species is already fairly common, we may be able to improve it considerably, in ways that decrease agriculture's reliance on nonrenewable resources.

If my summary of natural selection's strengths and weaknesses is accurate, then what I will call *tradeoff-blind biotechnology* (biotechnology need not be tradeoff-blind) and *misguided mimicry of nature* (not all mimicry of nature is misguided) have complementary weaknesses, both linked to tradeoffs.

Tradeoff-blind biotechnology mistakenly assumes that it will be relatively easy to improve plant traits that have already been improved by millennia of natural selection. In particular, biotechnology's oft-stated goals of improving water-use efficiency or photosynthesis are likely to prove elusive, as discussed in the next chapter.

With greater attention to tradeoffs, we may identify many opportunities for further genetic improvement of crop plants, using either traditional plant breeding or biotechnology, as discussed in chapter 8. But continually trying to reinvent nature's wheels is a recipe for continued failure.

Some advocates of copying nature, on the other hand, may be looking for wheels on rowboats. Landscape-scale features of even ancient natural ecosystems, like forests or prairies, haven't been tested competitively and are not necessarily an ideal model for agriculture. Some features may be worth copying, but considerable research will be needed to be sure. I will show that this conclusion strengthens, rather than undermines, the value of less-disturbed natural ecosystems as a source of ideas for agriculture.

Again, part of the problem is a tendency to ignore evolutionary tradeoffs. For natural ecosystems, the tradeoff is often between individual costs and collective benefits. For example, greater species diversity might make a forest less susceptible to disease epidemics. But natural selection will often favor individual-plant traits (like redwoods shading out shorter species) that undermine that goal. Therefore, *we should expect the adaptations of wild plants and animals to promote individual fitness more consistently than the overall organization of the natural ecosystems promotes efficiency or stability.*

Focusing on evolutionary tradeoffs can help us decide whether we should preserve sophisticated crop or livestock adaptations inherited from their wild ancestors, when we should reverse the effects of past natural selection that are inconsistent with our present goals, where we should seek new ideas among natural selection's innovations in wild species, and how we can slow harmful evolutionary trends in agricultural pests. But, whichever approaches we decide are most promising, let's hedge our bets by allocating some resources to alternative approaches.

# 5

## What Won't Work

TRADEOFF-BLIND BIOTECHNOLOGY

THIS CHAPTER DISCUSSES THE CHALLENGE of improving crop resource-use efficiency, using biotechnology or traditional plant breeding. I use the tradeoff principle from chapter 4 to argue that it will be difficult to improve efficiency more than natural selection already has, unless we pay more attention to tradeoffs. This hypothesis is supported by bio-technology's failure to deliver on past promises, relative to natural selection's successes. One human innovation missed by natural selection is discussed, as a possible counterexample. I close with a brief discussion of biotechnology's risks.

### Crop Genetic Improvement via Traditional Plant Breeding or Biotechnology

Plant breeding has traditionally relied on selection of visible or measurable traits. *Back-cross breeding*, which adds a small number of useful genes to an existing variety, has a long history of success.

For example, a wheat plant with high yield in the absence of disease might be crossed with (pollinating or pollinated by) another plant with genetic resistance to a particular disease. About 1/2 of the alleles in each resulting offspring will come from each parent. Those first-generation offspring may then be back-crossed to the high-yield parent, producing a second generation of plants with 3/4 of their alleles from the high-yield parent. Repeated back-crossing to the high-yield parent results in plants with more and more of their genome from that parent.

Only some of these plants will have inherited the key disease-resistance gene(s) from the disease-resistant parent. That's where selection becomes critical. By screening plants in each back-cross generation for disease resistance, and discarding those that fail this test, a plant breeder can produce a genotype with most of the desirable traits of the recurrent parent (for example, high yield) plus one or two desirable traits from the other parent.

Increasingly, plant breeders are using molecular information to guide this process. Consider breeding for apple fruit quality. It typically takes a few years before a tree is old enough to produce fruit to test. But if an allele for fruit quality has been identified, that allele will be present even in young seedlings. So it may be possible to select based on the presence of the fruit-quality allele, rather than on fruit quality itself. Testing to see whether a plant has a particular allele is now quick and easy, if we know the allele's DNA sequence. Even if we don't know exactly which allele(s) are responsible for high fruit quality, there may be another "marker" gene we can detect, near enough on the DNA strand to be inherited along with the fruit-quality gene. Because breeding guided by molecular information simply uses an additional source of information to select among plants from traditional, within-species crosses, this approach has raised few concerns among scientists and the general public.

A somewhat more controversial approach involves biotechnology—particularly, the direct insertion of genes into crop-plant genomes. The inserted gene may come from the same plant species, achieving results similar to what might be achieved by back-cross breeding, only faster. Or the gene may be from a completely unrelated species. A gene not found naturally in a plant species is often called a *transgene*, making the plant *transgenic*. For example, a bacterial gene that makes a caterpillar-killing protein has been transferred to several different crop species, making them resistant to caterpillars. Genetic engineering has also been used successfully to develop virus-resistant papaya.

"The potential of new technologies to change things for the better is invariably overstated, while the ways in which they will make things worse are usually unforeseen." That, at least, is the claim made in *The Victorian Internet*, a book about the telegraph.[96] At the end of this chapter, I will discuss ways in which some kinds of transgenic crops might make things worse. But what about potential benefits? Have they been overstated?

Biotechnology has already been successful at solving certain kinds of problems, including the development of pest-resistant crops. Of course, pest populations are evolving the ability to attack those crops, but not necessarily faster than if the pest-resistance came from traditional plant breeding. Biotechnology can also contribute to some longer-lasting improvements, including better nutritional quality. Such improvements can be longer lasting because they aren't countered by evolutionary changes beyond our control. For example, natural selection in wild sunflowers wasn't driven by the nutritional needs of humans (or birds), so it left considerable room for improvement (from our perspective) in the oil composition of sunflower seeds. I will argue, however, that some of the most-important benefits promised by the advocates of transgenic crops are unlikely to be achieved soon, if ever.

## Greater Resource-use Efficiency Is Needed

Population growth and other trends discussed in chapter 2 will require improvements in crop yields, not just better nutritional composition. Otherwise, food shortages and rising food prices will inevitably lead to more clearing of forests and draining of wetlands for agriculture, especially in countries with weak environmental laws. In some regions, food production can be increased by narrowing the *yield gap* (see glossary), bringing average farmer yields closer to what the best farmers already achieve. But raising average yields above 80 percent of potential is apparently quite difficult.[97]

*Yield potential* is defined as production per unit land area, without losses to pests or disease.[98] By this definition, an increase in pest resistance has no effect on yield potential. Some definitions also assume optimal soil fertility and/or water supply. Can transgenic approaches help increase yield potential? In chapter 8, I argue that this may be possible, but only if biotechnologists pay more attention to evolutionary tradeoffs. This chapter discusses the limitations of tradeoff-blind biotechnology.

Yield potential depends on fundamental crop growth processes, such as capture of light by photosynthesis, uptake of nutrients by roots, and efficient use of these resources to produce grain or other harvested products. It has often been claimed that genetically engineered crops will yield more, even with less water, because they will "make more efficient use of sunlight, water, and nutrients."[12] It is this specific claim whose validity will be the main focus of this chapter.

Resources like sunlight, water, and nutrients do not evolve as pest populations do, so improving their use is less of a moving target than improving pest resistance. Gazelles inherited their speed from ancestors who outran predators, but phosphorus molecules that escape capture by crop roots did not inherit this trick from elusive ancestral phosphorus molecules. Unlike pests, resources like photons, water molecules, and phosphorus molecules don't evolve to counter improvements in crops. Therefore, any genetic improvement in the efficiency with which plants use resources would provide much longer-lasting benefits than the temporary improvements in pest resistance that genetic engineering has achieved so far.

Unfortunately, tradeoff-free improvements in the efficiency with which crops use light, water, or nutrients are unlikely anytime soon. This is because these traits have already been improved by millions of years of natural selection. Any improvements that were missed by natural selection will often be too complex for today's genetic engineers to imagine or to implement. These claims will be supported first by theoretical arguments, and then by comparing natural selection's actual achievements with those of biotechnologists.

## Natural Selection Has Already Tested More Options Than Humans Ever Will

The entire history of plant breeding by humans, including the major improvements achieved by illiterate farmers before our modern understanding of genetics, is less than one-thousandth of the time over which natural selection operated on the wild ancestors of our crops. Natural selection operates worldwide, whereas selection by plant breeders is limited to small field plots. As Darwin noted, "Natural selection is daily and hourly scrutinizing, throughout the world, every variation, even the slightest; rejecting that which is bad, preserving and adding up all that is good."[26] ("Good," that is, by natural selection's criteria!)

In the previous chapter, I argued that past natural selection, operating worldwide for millions of years on the wild ancestors of our crops, is unlikely to have missed *simple, tradeoff-free* improvements.

If this argument is correct, that still leaves two classes of genetic improvement that may have been missed by natural selection: those that are complex and those that recognize and build on an understanding of tradeoffs. Unfortunately, many proposed contributions from biotechnology don't meet these criteria. Often, biotechnologists ignore tradeoffs, yet propose changes simple enough that they would already have evolved, if they consistently increased fitness.

Simple improvements, those unlikely to have missed by natural selection if they were tradeoff-free, include any change whose phenotypic effect *could have been* achieved with a simple mutation, even if biotechnologists use a more-complex approach. A change in a single DNA nucleotide can increase or decrease gene expression (for example, the amount of a particular enzyme that a plant makes), so increasing or decreasing gene expression qualifies as simple.

I don't mean to imply that increasing gene expression is necessarily easy, with current biotechnology. But that's my point: many phenotypes that are difficult for us to design and implement through genetic engineering must nonetheless have arisen repeatedly in the wild ancestors of our crops, through point mutations, insertions, deletions, or duplications. Plants with greater expression of a given gene competed against plants with less expression. Whichever alleles resulted in the greatest fitness (seed production, adjusted for any differences in seedling survival) took over the population. Today, if we simply increase expression of some key gene, we are almost certainly re-creating a phenotype that has already been rejected by past natural selection. That doesn't necessarily mean that we should reject that option, but we need to ask why it reduced fitness in past environments.

Similarly, what if a gene is turned on only under certain conditions? For example, some plants activate certain chemical defenses only when they or their upwind neighbors are under attack.[99] Mutants that turned these defenses on all the time must have arisen repeatedly, but they were apparently outcompeted. Biotechnologists who propose converting inducible defenses to constitutive ones should first ask why past natural selection rejected this approach.

It may seem that this discussion assumes that natural selection has tested all simple variants on existing genes. This is far from true, if "all simple variants" every DNA sequence that is similar to those that do exist. What I am really claiming is that mutants with increased expression of gene X have arisen repeatedly, not that every specific DNA sequence that would increase expression of gene X has arisen repeatedly.

Here is a simple mathematical example to illustrate the difference. One of the key photosynthesis proteins contains 489 amino acids. If we consider all possible proteins of that size, each position could be occupied by any one of 20 amino acids, so there are $20^{489}$ possibilities. If we assume a hundred million plants per 4 square kilometers, over 40 million square kilometers, for one hundred million years, we only get $10^{23}$ plants, a much smaller number. So it is mathematically impossible for every possible variant of this protein to have existed on earth. Combinations that have never existed have never been tested by natural selection, so only a tiny fraction of all possible proteins that size have been tested.

The fraction that has been tested is not random, however. As discussed earlier, evolution usually proceeds in small steps. What if we change one amino acid at a time? Then there are only 480 × 19 = 9291 possible one-amino-acid changes. Mutation rates are high enough[69, 70] that each hundred million plants is likely to include mutants with each of these simple changes, so natural selection has tested each of them under a wide range of conditions. Apparently, none of these variants worked as well as the current genotype.

Let me summarize my argument so far. If some trait that we imagine would be beneficial is not found in wild plants or animals, then either the trait is less-consistently beneficial (to individual fitness) than we imagine, probably due to tradeoffs, or else the trait is not simple enough to have evolved repeatedly by natural selection. The question is whether something too complex to have evolved by natural selection is simple enough to be designed and implemented by humans. In general, the answer appears to be "no," although there may be some exceptions. For example, humans have made little or no progress, so far, in improving photosynthesis, relative to what natural selection has already done. This is presumably because greater efficiency of this fundamental plant process, so critical to crop yield, consistently increased individual fitness in past environments.

## Yield Potential Has Increased Little in Recent Decades

I don't mean to suggest that biotechnology hasn't accomplished anything. Both plant breeding and biotechnology have been used to develop crops resistant to various pests. But to keep up with increasing population growth, we will also need to increase yield potential, the amount of useful product (usually grain) we can harvest from a crop protected from pests and weed competition. Some definitions of yield potential also assume plenty of water and nutrients,[98] but there are many areas where irrigation is not practical. So increasing yields under water-limited conditions is an important goal.

Average crop yields are still increasing in many countries, but there has been little recent progress in increasing yield potential. Improvements in average yields are often the result of more farmers gradually adopting the best methods and the best varieties. Much of the increase has involved improvements in pest control, but this approach has obvious limits, as protection can never exceed 100 percent.

Increases in average yield obscure the lack of progress in improving yield potential of major crops since about 1980. Agronomist Ken Cassman, who has worked at UC Davis, the International Rice Research Institute (in the Philippines), and the University of Nebraska, found little evidence for genetic improvement in yield potential of corn or rice from 1980 to 2000.[8] Mark Bell, at the International Center for Corn and Wheat, drew similar conclusions for wheat varieties developed in Mexico.[100] Despite plenty of technical progress in molecular methods since 1980, crop-yield potential appears to have stalled.

One way to measure improvements in yield potential is to grow today's crops side-by-side with older varieties. Such comparisons can be misleading, however. For example, plant breeders involved in the Green Revolution, which doubled or tripled yields of wheat and rice, released rice variety IR8 around 1966. IR8 had much higher yield potential than earlier varieties, but how much progress have we made since then?

When IR8 is grown side-by-side with newer varieties, the newer varieties have higher yield. But, by consulting old records, Shaobing Peng, a researcher working at the International Rice Research Institute, found that IR8 once yielded as well as the newer varieties do today.[101] One reason may be that pests have had longer to evolve the ability to attack IR8.

If we could somehow increase photosynthetic efficiency (photosynthesis per photon), that would increase yield potential permanently. This contrasts with gains from pest resistance, which will be undermined by ongoing evolution of pests. With more efficient photosynthesis, we could produce more food, thereby lowering food prices. Or we could grow the same amount of food on fewer acres of cropland, freeing land for parks and nature preserves. Similarly, an increase in water-use efficiency would let us grow more food while using less water.

But is improving either photosynthetic efficiency or water-use efficiency possible? Any tradeoff-free increase in photosynthetic efficiency would presumably have increased the fitness of plants in most past environments. This would be true even if plants sometimes have more photosynthate than they need.

As for water-use efficiency, our most important crops evolved in environments where water supply was sometimes limited. Even in wet climates, plants compete with neighbors for water on unusually hot or windy days. Because any tradeoff-free improvements in photosynthetic efficiency or water-use efficiency would have benefited the wild ancestors of our crops over most of their evolutionary history, we can safely assume that natural selection has been working on these two traits for a long time. Can we humans do better?

## Natural Selection Has Improved the Efficiency of Photosynthesis and Water Use

Before comparing biotechnology's promises to its actual progress, let us examine some examples of natural selection's past accomplishments. Is natural selection really as "immeasurably superior to man's feeble efforts" as Darwin claimed?[26] I will focus on "more efficient use of sunlight, water, and nutrients," a stated goal of biotechnology[12] with clear importance to yield potential.

Crops use sunlight for photosynthesis. This process is a perennial favorite for improvement by biotechnology. One popular target has been the elimination of photorespiration (see glossary), in which the key photosynthetic enzyme, *rubisco*, wastes energy by interacting with oxygen rather than carbon dioxide ($CO_2$).

There's an evolutionary explanation for this inefficiency. Most of the oxygen in the atmosphere was produced by photosynthesis. So when photosynthesis first evolved, there was little oxygen in the atmosphere and photorespiration was not a problem. Even today, with the oxygen concentration thousands of times higher than the $CO_2$ concentration, rubisco mostly takes up $CO_2$. The $CO_2$ is converted (in a series of steps) to sugars to fuel plant growth. Sometimes, however, rubisco mistakenly reacts with oxygen, reducing the efficiency of photosynthesis.

A second problem faced by both wild plants and crops is the direct link between photosynthesis and water use. Carbon dioxide from the atmosphere diffuses into leaves through pores called *stomata*. Rubisco inside the leaf then takes up the $CO_2$ using energy from sunlight. But all live cells must remain hydrated to maintain metabolic activity. Inside leaves, this inevitably results in water evaporation from wet cell surfaces and diffusion out through the same stomata. Crop plants lose much more water through this evaporation from inside leaves, known as *transpiration*, than they store internally in leaves or in fruits like melons. Most plants die if they dry out, so it is essential that transpiration not exceed water uptake from the soil. This is a problem when soils dry during drought.

Plants cope with drought in various ways. Some corn varieties roll their leaves so that they absorb less sunlight. This reduces heat buildup, decreasing water loss from transpiration. However, this reduced light interception also means less photosynthesis. Some plants drop leaves during drought, decreasing transpiration and photosynthesis even more. Simply closing stomata can reduce transpiration, but it also chokes off the $CO_2$ supply to the leaf interior, again reducing photosynthesis. So there are many ways to survive drought, but most of them reduce photosynthesis and therefore reduce yield.

In other words, drought tolerance (see glossary) doesn't usually increase a crop's ability to grow and produce grain with less water. We need crops that somehow photosynthesize more, but without using more water. Although plant breeders have recently made some progress on this problem, as discussed in chapter 8, natural selection had already found a way to increase water-use efficiency much more dramatically, while also eliminating photorespiration.

Natural selection's trick is to pump carbon dioxide into special compartments inside the leaf, and then carry out photosynthesis there. Inside these compartments, the $CO_2$ concentration is much higher than in the atmosphere. The higher $CO_2$ concentration makes rubisco interact mainly with $CO_2$ rather than oxygen, so photorespiration is negligible.

Why does this also increase water-use efficiency? Pumping $CO_2$ into photosynthetic compartments decreases $CO_2$ concentrations in the rest of the leaf interior, from where it was pumped. This lower $CO_2$ concentration pulls more $CO_2$ into the leaf from the atmosphere through stomata, just as a dry sponge soaks up more water. This means that the plant can close its stomata a bit more, reducing water loss, while still taking up plenty of $CO_2$ for photosynthesis.

This same trick, known as *C4 photosynthesis*, has evolved at least 31 times.[102] This repeated evolution was detected by mapping C4 photosynthesis on the plant family tree, or *phylogeny*. The way C4 photosynthesis is distributed across the phylogeny shows that it appeared independently in branches of 18 different plant families, sometimes more than once per family. Of our three major crops, corn uses C4 photosynthesis, but wheat and rice do not.

Cacti and some other plants, like the century plant, *Agave*, have evolved even greater water-use efficiency. These plants open their stomata at night to take up $CO_2$ for later sunlight-driven photosynthesis (with stomata closed) during daytime. They lose much less water through open stomata at night than they would during the daytime, when heating by the sun tends to increase transpiration. The main problem is that they can't store $CO_2$ gas itself. Instead, they use $CO_2$ to make an organic acid at night, from which they can release $CO_2$ the following day. There are limits to how much organic acid they can store, even in the thick stems and leaves characteristic of these plants. So they use water very efficiently, but their potential growth is less than if they took up $CO_2$ during the daytime. No problem—if water is abundant, *Agave* plants open their stomata and start taking up $CO_2$ during the daytime as well.[103]

## Biotechnology Has Not Improved Photosynthetic Efficiency

Natural selection's improvements to photosynthetic efficiency and water-use efficiency dwarf anything that plant breeders or biotechnology have achieved. Developing C4 photosynthesis in rice has been proposed, based on the idea that "polyphyletic [that is, repeated] evolution of the C4 pathway suggests that the transition from C3 to C4 is relatively simple."[104] So far, however, something that may have been simple for natural selection (given millions of years) has proved extremely difficult for humans. Making plants that switch photosynthetic systems depending on conditions, as *Agaves* do, would be an even bigger challenge.

But there is no shortage of promises. For more than 30 years, genetic engineers have been claiming that breakthroughs like more efficient photosynthesis or water use are imminent. Time and time again, they have failed to deliver crops with higher yield potential attributable to either. Let's examine efforts on photosynthesis first and then water use.

In 1975, a molecular biologist claimed to have plants with much lower photorespiration, leading to "an increased net photosynthesis of about 40 percent" and a forecast that substantially higher crop yield was imminent.[105] We are still waiting, more than 35 years later. In 1982, it was suggested that rubisco "may be beneficially modified in vitro and re-introduced into a plant."[106] Again, we are still waiting. Our ability to modify crop genes has certainly improved, but "beneficially" implies that we can design a more-efficient enzyme than that bequeathed to our crop plants by millions of years of natural selection. Nobody has done so yet and, as discussed earlier, they are not likely to do so anytime soon.

But they keep talking about it. In 1999, *Science* interviewed several biotechnologists about the prospects for improving crop photosynthesis.[107] T. John Andrews referred to rubisco as "nearly the world's worst, most incompetent enzyme," while William Ogren said rubisco was "not one of evolution's finest efforts." Despite this disparagement of natural selection, most of the ideas in the article for improving photosynthetic efficiency were more similar to industrial espionage than engineering. That is, they involved copying one or more of natural selection's inventions, rather than inventing anything new. Copying nature's best inventions is, of course, a theme of this book.

One of natural selection's innovations discussed in the 1999 *Science* article was a rubisco found in algae that is better than the rubisco in land plants at discriminating between $CO_2$ and oxygen. This reduces the problem of photorespiration. Some genetic engineers thought that we might improve photosynthesis by transferring this algal enzyme into crops. Murray Badger noted, however, that "the more specific and discriminatory a reaction becomes, the slower it gets." More recently, Andrews and colleagues have studied this apparent evolutionary tradeoff and concluded that "all rubiscos may be nearly perfectly adapted . . . optimizing this compromise between $CO_2/O_2$ specificity and the maximum rate of catalytic turnover."[4]

One recent analysis recognizes this tradeoff, but suggests that rubisco may be better adapted to past conditions than to present ones.[27] As atmospheric $CO_2$ has increased, natural and human-imposed selection have apparently lagged behind. Assuming a simple tradeoff curve, it appears that rubisco in today's crops is adapted to the $CO_2$ levels to which its preindustrial ancestors were exposed, about half of today's levels. With today's higher $CO_2$, specificity is less important, so we might be able to trade some of that specificity for a faster reaction rate.[27] This is essentially the opposite of what biotechnologists were suggesting in 1999, reflecting an encouraging increase in attention to tradeoffs.

Another idea discussed in 1999 that remains popular today is to copy C4 photosynthesis from corn into rice. (Another evolutionary tradeoff is that C4 photosynthesis performs poorly in colder climates, so tropical rice would be a better target than temperate wheat.) One group claimed to have gotten higher photosynthesis and up to 35 percent higher yield, just by transferring a few genes from corn into rice.[108] These higher yields have not been replicated independently, however—a recurrent problem for claims made by biotechnology. In fact, the higher the expression level of these alien genes, the *lower* the photosynthesis.[109] So far, copying natural selection's innovations through biotechnology has proved difficult.

A recent analysis by molecular biologist Stephen Long and colleagues concluded that these "most widely promoted strategies—conversion of C3 crops to C4 and improved specificity of rubisco—may be the most difficult to achieve and, from a theoretical basis, might result in lower and not higher" photosynthetic efficiency.[110] Instead, they suggest approaches like optimizing leaf angle.

I agree that changing leaf angle is an example of a more-promising approach, and biotechnology might offer a faster route to such changes. But conventional corn breeders have been releasing varieties with increasingly vertical leaves since about 1960, apparently as a side-effect of selecting for yield.[84] Experiments showing potential benefits of erect leaves were done long before the biotechnology revolution.[111] There are tradeoffs, of course, including a need for more plants per acre. Leaf angle and other tradeoff-linked traits were discussed in an evolutionary context by the Australian agronomist, Colin Donald, back in 1968.[112] Donald's ideas, which were a major inspiration for this book, will be discussed in chapter 8.

## Tradeoffs Limit Biotech Improvement of Crop Water Use

Genetic engineers have also repeatedly promised improvements in *drought tolerance* (the ability to survive drought) and/or *water-use efficiency* (actually producing more harvestable produce using less water). So far, there is no clear evidence that they have made tradeoff-free improvements in either. Breakthroughs are repeatedly described as imminent, but none has yet stood up to independent testing under relevant field conditions.

In 2008, two leading general-science journals, *Science* and *Nature*, both had feature stories about the prospects for increasing crop yields with limited water. Roger Beachy, then working at a research center linked to a major agricultural biotech company, admitted that biotechnology advocates had made "some remarkable and extravagant predictions back in the 1980s . . . [and] it comes back to haunt you."[113] The "haunting," however, seems to come in the form of occasional criticism, rather than any actual consequences, like reduced funding. Beachy, a strong advocate of biotechnology, was subsequently put in charge of the U.S. Department of Agriculture's entire research program.

In one of those 2008 articles, Marianne Bänziger, working in Mexico at the International Maize and Wheat Improvement Center, noted that "there are probably hundreds of groups that work on transgenic drought tolerance . . . but very few have made it into the field and shown yield increases."[113] No specific examples of successes were given. Pasquale Steduto was even more blunt, saying that "so far, we do not have a direct gain from GM or molecular biology in terms of drought resistance."

But such articles always hold out the prospect of a breakthrough on the horizon. I am reminded of the saying that "nuclear fusion is the energy source of the future, and it always will be." Both *Nature* and *Science* referred to the same recent paper in *Proceedings of the National Academy of Science*.[114] The authors, all affiliated with the same biotech company, increased the expression of an existing gene (in this case, one that controls other genes) and reported improved performance of corn under drought. This result, if confirmed, might seem to disprove the hypothesis that prolonged natural selection rarely misses simple, tradeoff-free improvements, because increasing the expression of an existing gene meets my definition of a *simple* genetic change.

But was this really a tradeoff-free improvement? Under drought, the genetically engineered corn reportedly had "improved yield ($P < 0.1$)." The statistical term $P < 0.1$ means that there is less than a 10 percent chance that the apparent yield difference was just coincidence. Most scientific journals, including all mainstream agricultural ones, insist on a higher standard, less than a 5 percent chance that the results are due to chance. This standard is intended to reduce the chances of drawing incorrect conclusions if, for example, the transgenic plants just happened to be assigned to plots with better soil.

But my concerns go beyond statistics. As discussed earlier, mutations that increase gene expression happen all the time, and natural selection retains those that are beneficial to the plant. So why does corn normally have lower expression of this gene than was obtained by genetic engineering?

Could greater expression of this gene be harmful under some conditions? For example, could the genetically engineered corn have lower yield than existing varieties, given normal or above-normal rainfall? Mainstream agricultural journals usually insist on tests over at least two years or in at least two locations with different weather, to detect such tradeoffs if they exist. I would have expected authors from a university or other nonprofit research institute to include yield data for a normal year in their paper. Although the biotech-company researchers did apparently grow some plants with "ample water supply," they did not report yield data for those well-watered conditions. I wonder why.

The significance of their claimed accomplishment could easily be resolved by comparing their "drought-tolerant" corn with the best currently available corn varieties, under a range of conditions. Obviously, this comparison would have to be done by independent researchers, without any financial incentive to favor one variety over another.

Poor performance under well-watered conditions is not the only possible tradeoff associated with drought tolerance. For example, there could be a tradeoff between individual-plant competitiveness and drought tolerance. Some such tradeoffs may be acceptable, as discussed in chapter 8. But what has been published so far is not convincing evidence that this highly touted "breakthrough" is a genuine example of a tradeoff-free improvement.

## Would Genetic Engineering of Nutrient-use Efficiency Help?

Increasing nutrient-use efficiency seems like a good idea, but it might not have as much real impact as increased water-use efficiency. A key difference is that most of the water taken up by a crop is transpired to the atmosphere. Very little of the water used by a crop ends up in the grain itself, so it's theoretically possible to use less water and still produce the same amount of grain. This contrasts with nutrients like nitrogen, an essential ingredient in grain protein. Typically, 70 percent or more of the nitrogen taken up by the crop ends up in grain protein.[115] Therefore, a 50 percent decrease in nitrogen uptake would almost inevitably decrease grain protein, due to conservation of matter.

One widely used definition of *nutrient-use efficiency* is the ratio of a plant's *growth* (total weight at harvest) to its *nutrient use* (usually defined in practice as its nutrient content at harvest). High nutrient-use efficiency, so defined, is not necessarily desirable.

To see why, consider nitrogen-use efficiency. Plants have various adaptations that maintain growth in the face of nitrogen deficiency. Therefore, reducing a plant's nitrogen uptake by half will reduce growth by less than half. This means that a nitrogen-starved plant will automatically have lower percent nitrogen, giving it higher nutrient-use efficiency, as just defined.

So, if nitrogen-use efficiency were our only goal, we could simply breed for defective roots! They would take up less nitrogen and end up with lower percent nitrogen—that is, greater nitrogen-use efficiency. They would, however, have lower yield (less seed or lower-protein seed) than plants with better roots growing in the same soil. Can we do better than this?

Genetic engineers have suggested that crops with a more-efficient rubisco—let's temporarily assume that natural selection somehow missed this improvement—would need less rubisco. Because rubisco is a protein, it contains nitrogen, and because it is so abundant, it can contain a large fraction of all the nitrogen in a leaf. So, the reasoning goes, a crop with more-efficient rubisco (more photosynthesis per gram of rubisco) would need less nitrogen.[107]

Plants with more-efficient rubisco might indeed require less leaf nitrogen for the same photosynthesis rate. However, this would probably not reduce yearly nitrogen needs. As noted earlier, most of the nitrogen in rubisco is eventually recycled into seed proteins, at the end of the growing season. With less nitrogen available from recycling less rubisco, crops would produce less grain or lower-protein grain.

Corn already has the nitrogen-efficient C4 photosynthesis system, so it can photosynthesize with less rubisco and therefore less leaf nitrogen than some other crops. But as soon as it starts making seeds, it needs nitrogen for grain protein. Corn's high seed yield puts its nitrogen needs above those of many other crop species, despite its nitrogen-efficient photosynthesis. Conservation of matter strikes again.

The same argument applies to efficiency in using other nutrients. Nutrient-use efficiency needs to be examined in the context of whole-farm nutrient budgets. Crops that take up nutrients faster may reduce some wasteful nutrient losses (for example, leaching [see glossary] of nutrients in water percolating down through the soil), but every atom of nitrogen or phosphorus that is sold off-farm in grain, milk, or other farm products still needs to be replaced, for long-term sustainability.[37] Increasing the fraction of total crop nitrogen that ends up in grain (that is, the *nitrogen harvest index*) beyond its current 70 percent or more could be worthwhile. Other than that, changing the way crops use nitrogen after they take it up doesn't seem to have much potential. Root systems that reduce erosion would probably have a bigger impact.

So far, we have no evidence—at least, no evidence confirmed independently—of any biotech crop with "more efficient use of sunlight, water, and nutrients,"[12] except perhaps as a side-effect of insect or disease resistance. This record of failure contrasts with natural selection's successes. These successes include the repeated evolution of C4 photosynthesis, with its more efficient use of sunlight (by eliminating photorespiration) and water (by taking up more $CO_2$ through partially closed stomata, thereby reducing transpiration), plus its minor contributions to nitrogen-use efficiency.

## Near-term Benefits of Biotechnology Have Been Exaggerated

So far, biotechnology has done little or nothing to increase yield potential, but what about actual yield? For example, have improvements in pest control been significant?

The Union of Concerned Scientists published a study of actual yield gains from genetic engineering.[116] Consistent with the discussion earlier, they found that "genetic engineering has not increased intrinsic yield," what most agronomists call *yield potential*. This was my main point. But they go further.

They found that herbicide resistance, one of the two most widely used transgenic traits, has not increased yield of either soybean or corn. (Herbicide resistance is popular with farmers anyway, because crops that can survive herbicide sprays have lower weed-control costs, at least until weeds evolve resistance to the same herbicides.) The other widely used trait is insect resistance based on the *Bacillus thuringensis* (Bt) toxin (see glossary), derived from a bacterial gene. For Bt crops, they found yield increases up to 12 percent, relative to conventional nontransgenic varieties . . . but only where pest populations are high. Averaging over locations varying in pest populations, they estimate a yield gain of 2 percent. Their analysis wasn't published in a scientific journal, where it would have been subject to the quality control of peer review, so maybe this estimate will turn out to be too low. But 2 percent is an awfully small improvement, relative to 40 to 60 percent increase in yields of major grain crops needed between 2000 and 2030.[8]

Decades of unfulfilled promises from biotechnology advocates and practitioners bear a striking resemblance to vaporware. *Vaporware* is a disparaging term for the exaggerated claims that software companies sometimes make about products that have not yet been released. The original idea, supposedly, was to discourage competitors so much that they abandoned development of competing products. One biotechnology journal published a paper titled "Vapornomics," perhaps implying deliberate deception similar to that practiced by software companies.[117] Sincere overenthusiasm may be more common than outright deception, however.

What about future prospects? I have argued that simple, tradeoff-free improvements in traits key to crop yield potential (particularly the efficiency with which crops use sunlight, water, and nutrients) are unlikely to have been missed by past natural selection. My analysis of progress to date is consistent with this hypothesis. We have no clear examples of tradeoff-free improvements in efficiency from simple genetic changes, such as increasing the expression of an existing gene or changing a few amino acids at the active site of an enzyme.

Fortunately, my hypothesis does not completely exclude the possibility of major genetic improvements in crop yield potential. The two most promising routes to improving yield potential are those involving tradeoffs that we are willing to accept (discussed in chapter 8) and improvements so complex that they might have been missed by natural selection, despite their benefits. I will briefly discuss the latter approach here.

## What Complex, Tradeoff-free Improvements Have Been Missed by Evolution?

What are the prospects for truly novel improvements in photosynthesis through genetic engineering? As discussed earlier, we cannot assume that natural selection has tested every possible genotype, or even every possible phenotype. However, natural selection probably has tested every phenotype that is reasonably similar to those that already exist.

The evolution of C4 photosynthesis from C3 photosynthesis is a convenient standard of complexity. Improvements this complex are apparently something that natural selection can do fairly easily, given that C4 photosynthesis has evolved at least 31 times,[102] yet such improvements may be beyond anything humans today could design and implement from scratch. Even transferring the existing C4 system has been beyond our abilities, at least so far. It's not just a matter of an enzyme or two. We also need compartments within the leaf where $CO_2$ can be concentrated, and some way of pumping $CO_2$ in without having it leak out.

Designing new enzymes or whole photosynthetic systems from scratch would be real *genetic engineering*. Most of what we have achieved so far is tinkering: making minor changes in genes that evolved by natural selection, increasing their expression, or moving them around. Tinkering is also a reasonable description of what natural selection does,[118] but natural selection has been tinkering for millions of years, testing trillions of plants at once. As long as we are limited to genetic tinkering, we are unlikely to discover phenotypes much different from those already tested by natural selection. How soon will we be able to move from genetic tinkering to genetic engineering?

One recent example suggests that true genetic engineering may be possible sooner than I would have expected. Rashad Kebeish and colleagues have invented a partial solution (apparently missed by natural selection) to the problem of photorespiration.[119] Photorespiration was briefly described earlier, but now we need a little more detail. When the photosynthetic enzyme, rubisco, takes up $CO_2$, the first product is P-glycerate, which eventually gets processed into useful molecules like sugars. But when rubisco mistakenly interacts with oxygen instead of $CO_2$, the product is a fairly useless molecule, P-glycolate.

C4-photosynthesis plants solve this problem by keeping rubisco in a compartment where $CO_2$ concentration is so high that rubisco makes very little P-glycolate. The more common C3-photosynthesis plants, on the other hand, have to process the P-glycolate through a complex series of steps, eventually recycling its carbon molecules back into $CO_2$. Rubisco then gets a second chance to take up that $CO_2$.

The problem is that the $CO_2$ is released in the mitochondria. These organelles, found inside all plant and animal cells, are the normal site of respiration, where oxygen and sugars (or other energy-rich molecules) are converted to $CO_2$ and usable energy. Mitochondria and chloroplasts (see glossary), the organelles where photosynthetic $CO_2$ uptake occurs, are the distant descendants of symbiotic bacteria that colonized (or were captured by) cells over a billion years ago.[75,76] Kebeish and colleagues knew that any $CO_2$ released in the mitochondria might escape before encountering a rubisco molecule in a chloroplast. If the $CO_2$ recycled from P-glycolate could be released in the chloroplast instead, that would increase the $CO_2$ concentration around rubisco, reducing subsequent photorespiration. So they moved $CO_2$ recycling to chloroplasts. To do this, they added five bacterial genes to the chloroplasts of *Arabidopsis*, a small weedy plant often used for testing biotech ideas.

The resulting transgenic plants grew significantly bigger than control plants. But was this difference really due to more-efficient recycling of $CO_2$? To test the recycling hypothesis, they grew the plants under higher $CO_2$ concentrations; this should eliminate the advantage of better $CO_2$ recycling. Sure enough, the growth difference disappeared under high $CO_2$.

This *Arabidopsis* work has not yet been repeated with crop plants, where adding multiple genes from other species can be more difficult. So we don't know whether a similar approach would increase crop yield significantly and whether there will be any significant tradeoffs. One problem with complex improvements is our limited ability to predict such tradeoffs in advance.

If this does turn out to be a tradeoff-free improvement, which seems possible, why it was missed by natural selection? Earlier, I proposed C4 photosynthesis, which natural selection has invented repeatedly, as a standard for measuring the complexity of an improvement. One could argue that moving $CO_2$ recycling to the chloroplasts is no more complex than C4 photosynthesis, yet it appears to have been beyond the reach of natural selection. But remember that my definition of *simple* includes improvements requiring a series of steps, so long as each step increases fitness. Maybe C4 photosynthesis, despite its current complexity, was achievable through a series of fitness-improving steps, whereas there was no such pathway to move $CO_2$ recycling from mitochondria to chloroplasts.

So this example isn't necessarily a case of natural selection missing a *simple*, tradeoff-free improvement. But it may show that human creativity can sometimes find *sophisticated* solutions to problems too difficult for natural selection. Let's see whether these laboratory results with *Arabidopsis* actually lead to more-efficient crops.

## Genetically Engineered Crops Have Some Risks

The potential risks of biotechnology—real and imaginary—have received much more attention than its failure to deliver on some of its promises. So I won't devote too much space to this issue. Actual and potential benefits depend on the particular trait being altered. This is equally true of potential risks. The trait-specificity of possible risks implies a need for effective, science-based regulation. Wholesale bans or labeling requirements for food products derived from transgenic crops would be appropriate only if all such food posed similar health threats. Similarly, wholesale bans on the growth of transgenic crop plants (even if not sold for food) would be appropriate only if they all posed similar environmental risks, such as endangering wildlife.

But most risks are crop- or trait-specific. Some foods derived from transgenic crops might be unhealthy for particular groups. The potential transfer of allergen genes (for example, from peanut into corn) would be an example. Because such direct health risks are unlikely to be ignored by regulatory agencies, I will let them worry about it.

I am more concerned about health risks from transgenic crops that are intended not for human consumption, but for industrial purposes. For example, crops that produce pharmaceuticals could reduce the costs of life-saving drugs. But one tomato looks much like another. Drug-producing genes could be carried to garden tomatoes by pollinating bees that somehow make it through corporate security checkpoints. Or seeds could be stolen and end up in the underground economy. Once tomatoes or other food crops are transformed to make drugs, this sort of accident seems inevitable. What if drug-producing tomatoes grown by an AIDS patient for her own use are stolen by a greedy neighbor, who consumes a higher, toxic dose? Here's one possible solution: don't put genes for drug production in food crops![120] There are lots of crops, from tobacco to jojoba, that nobody would eat by mistake.

Biotechnology can also be used to eliminate toxic chemicals from crops, rather than adding them. This could certainly be a positive development, but there are some risks as well. For example, cassava root, an important food in some tropical countries, often contains high levels of cyanide. Cyanide production in cassava is similar in mechanism and function to cyanide production in clover, discussed previously. High-cyanide cassava varieties require special treatment, using traditional methods like boiling and drying, before they are safe to eat. It has been suggested that biotechnology could be used to reduce or eliminate cyanide in cassava. Low-cyanide roots might be more susceptible to various pests, but would it be a good idea, apart from that?

What happens if all the farmers in one region switch to the low-cyanide varieties, but those elsewhere do not? People in the low-cyanide region could skip the laborious process of detoxification, but what if some out-of-town cassava finds its way into local markets? Like drug-producing tomatoes, high-cyanide roots wouldn't necessarily look any different.

It turns out that nonhuman pests are not the only problems solved by cyanide in cassava. Some subsistence farmers prefer high-cyanide cassava because it is less likely to be stolen, even during famines.[121] They grow low-cyanide varieties to sell and high-cyanide varieties as starvation insurance. High-cyanide roots can be left in the ground for a year or more until needed, then dug up and processed as food. Also, cyanide contains nitrogen and it can be an important nitrogen source for growth and protein production in roots. So as an alternative to eliminating the cyanide, plant geneticists Dimuth Siritunga and Richard Sayre used molecular methods to develop a cassava variety in which cyanide is still present, but is eliminated much faster during processing.[121] This innovative approach combines some of the advantages of low- and high-cyanide cassava.

Benefits and risks to the environment from growing transgenic crops, like health benefits and risks from eating them, depend on crop-specific details. Millions of acres of transgenic crops are grown in the United States and other countries, but two key traits predominate, at present. Crops resistant to herbicides, like glyphosate, make it easier to control weeds. They reduce the need for tillage operations that can increase soil loss to erosion. Also widespread are transgenic crops that make an insecticide derived from the bacterium *Bacillus thuringiensis* (Bt), which mainly kills insect pests that eat the crops. This built-in insecticide can replace sprayed insecticides that are dangerous to beneficial insects, wildlife, and humans, especially those living in agricultural areas. Pests will evolve resistance to pest-resistant transgenic crops, a problem discussed in chapter 10. But when that happens, will we be any worse off than before those crops were developed?

One possible problem is that reliance on transgenic pest- or herbicide-resistant crops may lead agricultural researchers to neglect other approaches. Then, when these crops fail, we may have fewer new control methods in the pipeline. This could be due to lack of funding for nonbiotech approaches, as well as overall complacency.

Evolutionary risks associated with herbicide-resistant crops have received less attention than those linked to Bt crops. One obvious concern is more rapid evolution of herbicide resistance in weeds. Herbicide-resistant weeds are not usually "superweeds" in any other respect, but they can still be a serious problem. Herbicide resistance can evolve with or without gene flow from crops to weeds. As discussed in chapter 3, gene flow can happen when pollen from a crop fertilizes a nearby weed, leading to a crop-weed hybrid.

Gene flow is possible only when the weeds are related to the crops, but this situation is more common than you might think.[122] For example, johnsongrass, a serious weed, hybridizes with sorghum.[123] If all first-generation crop-weed hybrids died without producing seed or pollen, then gene flow would have no long-term effect. However, some hybrids may survive, effectively adding crop genes to the weed's gene pool. A sunflower crop, grown only once, left alleles behind in a wild sunflower population that could still be detected five years later.[68] Fortunately, the crop wasn't herbicide resistant, so farmers didn't end up with herbicide-resistant wild sunflowers shading out their crops.

Although much of the controversy over gene flow from crops has focused on transgenic crops, gene transfer from nontransgenic crops is also a potential concern. Plant breeders sometimes use the wild relatives of crops as sources of useful genes, in either traditional or high-tech breeding programs. Wild plants growing near genetically related crops can have their gene pools diluted by gene flow from crops. As discussed earlier, I don't think we need to worry that genetic engineers will soon develop drought-resistance genes that would spread to weeds. However, a gene for resistance to herbicides or perhaps disease might increase the survival and seed production of crop-weed hybrids, relative to the wild parents. Half of the alleles in any first-generation hybrid come from each parent, so potentially valuable alleles might be lost from wild populations in this way.

Even without gene flow, greater use of herbicide-resistant crops may speed the evolution of herbicide-resistant weeds, if they lead to excessive reliance on one herbicide, thereby increasing the intensity of selection (see glossary), as discussed in chapter 10. Eleven weed species have already evolved resistance to glyphosate herbicides.[124] Usually, this occurs in places where glyphosate is used as the primary method of weed control over a period of years.[125, 126] Herbicide-resistant crops make over-reliance on one herbicide more likely, but they are not solely responsible for this practice.

I have not considered all of the possible environmental effects of Bt crops or herbicide-resistant ones, nor possible risks of other transgenic crops or livestock that might be developed in the future. Nor, with some important exceptions,[122,127,128] have many other people. Would genetically engineered chickens be more or less likely to incubate influenza viruses that could also infect humans? Is the availability of Bt corn reducing the use of practices like crop rotation to control insect pests? If so, will abandonment of these practices lead to other problems, such as a buildup in soil-borne pathogens? Remember the 80-year-old experiment at Rothamsted that had to be abandoned because of a particularly persistent soil pathogen?[47] Advocating more attention to unexplored risks does not imply that risks necessarily exceed benefits.

## Should "Naturalness" Be a Key Goal?

I was once invited to a biotech-industry-sponsored meeting on "The Ethics of Biotechnology," focusing mainly on whether biotechnology is "natural." Similarly, it has been claimed that the "commonest reason the public gives for buying organic food is that it is natural."[25] Arguing about whether something is "natural" only seems worthwhile if we were already in agreement that *natural* and *good* are synonyms. One of the themes of this book is that this is not always true.

Certainly, the more we depart from nature, the more we enter unexplored territory, with possible unknown risks. But should we worry about whether bicycles are natural, or should we focus on real safety issues: lights and reflectors, helmets, and bike lanes?

Actually, bicycling *was* once denounced as unnatural, at least for women. In 1896, pioneering feminist Susan B. Anthony wrote that: "Bicycling . . . has done more to emancipate women than anything else in the world."[129] The eventual societal effects of the increased freedom of movement bicycling gave women is an example of how the effects of new technology, positive or negative, can be hard to predict.

I doubt that arguments about whether biotechnology is "unnatural" or intrinsically "unethical" (because of conflicts with particular religious beliefs, for example) will lead to a consensus upon which we can base sensible laws and policies. A Swiss law protecting the "dignity" of plants has been applied to biotechnology,[130] but there is no agreement on how dignity is defined for species without brains. When traditional plant breeders choose only a few plants as parents for the next generation, is that a grave insult to the rest?

## Perspective

The risks of biotechnology, though real, seem small relative to the potential benefits promised by biotech advocates. The question is, how many of those promised benefits will actually materialize?

In evaluating the risks and benefits of transgenic crops, let's distinguish between short- and long-term effects. Pest- and herbicide-resistant crops will provide only a short-term benefit, because pests and weeds will continue to evolve. On the other hand, crops with greater photosynthetic efficiency or water-use efficiency would reduce agriculture's needs for land or water forever, no matter how weeds and pests evolve. Biotechnology has not yet led to such long-term improvements, however. Nor do such improvements seem imminent, although the $CO_2$-recycling work in *Arabidopsis* may suggest a promising approach. But the biggest opportunities, I will argue, require greater attention to tradeoffs.

Before turning to tradeoff-linked opportunities that might benefit from biotechnology, however, let's consider innovations inspired by patterns seen in natural ecosystems. This approach is often perceived as diametrically opposed to biotechnology, yet it has an equally critical need for an evolutionary perspective.

# 6

## Selfish Genes, Sophisticated Plants, and Haphazard Ecosystems

OUR SECOND core principle is that competitive testing is more rigorous than testing merely by persistence. This chapter uses that principle, and actual data, to argue that the overall organization of even ancient natural ecosystems may be relatively imperfect, compared to individual adaptations that have been repeatedly tested through competition. Natural ecosystems are still important, however, as the context where sophisticated adaptations of wild species evolved.

### How Perfect Are Natural Ecosystems?

Aesthetics alone can justify preserving natural ecosystems. I would usually rather visit a natural prairie than a corn field, or a natural forest instead of an orchard. Returning to the airport from our honeymoon on Kauai, my wife and I were surrounded by people talking about the great time they had there . . . playing golf. I'm just glad they weren't all out in the Alakai Swamp with us. We didn't see another human all day, just a breathtaking variety of plants, birds, and one fascinating spider.

There are many other reasons to preserve natural ecosystems. Because so much of their value is intangible, economic arguments can't ever justify destroying a natural ecosystem. But economic arguments can often justify preserving them. For example, they provide many economically valuable ecosystem services, such as purifying air and water.[57]

A major theme of this book is that natural ecosystems also have enormous economic value as a source of ideas for improving agriculture. But how do we identify the best ideas? That's the tricky part.

I argued in the previous chapter that natural selection has repeatedly tested most of the simple variations on the ways individual plants acquire and use resources like light and water. Therefore, we may want to copy the individual adaptations of wild species, which will often be more sophisticated than anything we could develop through biotech tinkering.

In seeking nature's wisdom, should we also copy the overall organization of natural ecosystems? By overall organization, I mean things like the total number of species and their relative numbers. For example, a natural meadow might have mostly nitrogen-fixing legumes, mostly fast-growing grasses (which can't use atmospheric nitrogen), or anything in between. Is the ratio in a natural meadow likely to be optimal, in terms of sustainability or productivity?

Spatial arrangements of species (for example, whether plants of the same species are clustered together or dispersed) are another example. Should we expect to see the same degree of sophistication in the spatial patterns of trees in forests that we do in the spatial patterns of leaves on trees? That's the central question for this chapter: *Have ecosystem-level features, such as which species are present and how they are distributed in time and space, been reliably improved by natural processes?*

What if the answer is "no"? Should we still study ecosystem organization and how it affects ecosystem function? Absolutely—we might learn something useful. But should we *copy* the organization of natural ecosystems in designing agricultural ecosystems? Only sometimes.

Why not always, or at least usually? If some natural process had consistently improved the organization of natural ecosystems over millennia, in ways that improve ecosystem function, then that organization might often be better than anything we could design ourselves. In particular, agricultural landscapes that copy the organization of continuously improved natural ecosystems might be more sustainable and efficient in the use of scarce resources.

But I will argue that this is not generally true. Although mere persistence is evidence of sustainability, at least under past conditions, it is not evidence of consistent improvement over millennia. Forests have not been improved by competitive testing against each other, as individual trees have. If natural selection (or some other natural process) hasn't consistently tested and improved ecosystem organization, then we will need to do our own tests before deciding whether the number of species, the legume:grass ratio, or the spatial patterns we see in natural ecosystems are better than alternative numbers, ratios, or patterns.

When I ask whether any process has consistently "improved" (or perhaps perfected) the overall organization of natural ecosystems, I mean by criteria relevant to agriculture. Those criteria, discussed in chapter 2, include productivity (to use no more land than necessary), efficiency in the use of scarce resources (to leave some water in rivers, for example), stability over years (to prevent even occasional famines), and sustainability (to maintain all of these benefits over the long term).

Given our focus on agriculture, we don't need to determine whether natural processes have consistently improved the organization of natural ecosystems by other criteria, such as increasing biodiversity. *Natural* biodiversity may be an end in itself—we like flowers and birds—but *crop* diversity matters mostly because of its contributions to productivity, resource-use efficiency, risk-reduction, and sustainability, plus any aesthetic or wildlife-related benefits. (Remember, productivity and resource-use efficiency are key to limiting agriculture's demand for more land or water, resources upon which the biodiversity of natural ecosystems depends.) Even if we found that increasing biodiversity actually *decreased* the productivity of natural ecosystems in some bioregion, we might still want to preserve that biodiversity for aesthetic reasons, but we wouldn't necessarily copy that level of diversity on our farms.

If natural ecosystems haven't been tested through repeated competition with alternative ecosystems, are there other natural processes that may have improved landscape-level organization, in ways that consistently increase productivity, efficiency, and sustainability? If not, then we shouldn't simply copy natural ecosystems. But they would still be worth studying as a possible source of information on what works—and what doesn't. For example, we might assess how species diversity affects stability, based on actual data on wildlife population cycles in forests with higher and lower diversity. But we can't just *assume* that farms that copy the organization of natural ecosystems would have greater year-to-year stability, unless we find evidence that natural processes have actually adjusted natural-ecosystem organization in ways that consistently improve stability.

So has some process improved natural ecosystems over millennia, by criteria relevant to agriculture? If the Alakai Swamp had one fewer orchid species, would it be a worse ecosystem? Well, it would have one fewer orchid species, and we like orchids. But what if it had one additional orchid species, relative to current numbers? Would that additional species make the ecosystem worse at absorbing heavy rain without losing too much topsoil? Would an extra orchid make the ecosystem provide fewer wild pollinators for nearby farms? If we assume that natural ecosystems have somehow been perfected, then a change in species diversity in either direction should be a change for the worse. This seems unlikely.

I assume that some of the ways we might make an orchard more like a natural forest, or a pasture more like a prairie, would be improvements. But can we say more than that? If some process has *consistently* improved the organization of natural ecosystems, by criteria relevant to agriculture, then *any* change to greater "naturalness" would probably be an improvement. But I suggest that any improvement of ecosystem structure over millennia has been much less consistent than the improvement of individual adaptations. Therefore, each time we consider copying some landscape-scale feature of a natural ecosystem, we will need to test it. And we will need to test it more thoroughly than if we were copying some individual adaptation, already tested by natural selection's process of competition against various alternatives. We do have to keep in mind, however, that natural selection's criteria can differ from our own.

## Agroecology and the Assumed Perfection of Natural Ecosystems

The near-perfection of natural ecosystems is apparently the foundational hypothesis for many scientists who call themselves *agroecologists*. For example, a paper by a well-known member of this group claimed that each wild species has "evolved so that it does not overexploit its resources."[131] If this were actually true, that could make natural ecosystems a model of stability and sustainability. But see the discussion of population crashes in wild caribou, later in this chapter.

Another paper claimed that "managing undesired variables in agricultural systems is similar to that for other systems, including the human body."[132] Human bodies have various homeostatic (stabilizing) mechanisms (see glossary) and an immune system, both improved by millennia of natural selection. Do the landscape-level homeostatic and disease-limiting mechanisms found in ancient natural ecosystems have a similar history of improvement? If so, by what mechanisms?

The assumed perfection of natural ecosystems has led to the idea that we should copy them, to design better agricultural ecosystems. One leading agroecologist made this point explicitly, writing that "natural ecosystems can be regarded as models for pest management strategies and agroecosystems."[132] This theme is echoed in the title of another book, *Agriculture as a Mimic of Natural Ecosystems*,[133] although contributors to this edited volume actually held varying views.

An influential paper with a similar theme will be the main focus of the next chapter, as its perspective is representative of many agroecologists. That paper noted that "natural plant and animal communities have been shaped by climatic and evolutionary histories beyond complete human comprehension."[14] The authors went on to claim that, nonetheless, "patterns and processes discernible in natural ecosystems still remain the most appropriate standard available to sustainable agriculture." So they suggested that we can't understand natural ecosystems, at least not thoroughly, but we should copy them anyway. Does that make sense?

Ideally, we would like to understand something before copying it. Physicist Richard Feynman wrote about "cargo cults" on remote Pacific islands after World War II, whose members tried to attract renewed visits from cargo-laden airplanes by making imitation runways, complete with fake radio huts with bamboo antennas. Real radios might bring in an airplane, but a real-looking copy will not. Feynman's concern was "cargo-cult science,"[134] like scientists using expensive equipment to create the illusion that they are doing cutting-edge research. Is copying aspects of nature that we don't understand "cargo-cult agroecology"?

Not always, perhaps. If we are fairly certain that some aspect of nature has been repeatedly tested, and improved, by our criteria, then it may make sense to copy it, even without complete understanding. For example, plant breeders have long "copied" disease-resistance genes into our crops, by crossing them with their wild relatives. Until recently, we had no way to analyze those copied genes in detail, but we did know that they had been repeatedly tested, in competition with alternatives, by past natural selection.

By analogy, could it make sense to copy the organization of natural ecosystems (species diversity, for example, or spatial patterns), even if their effects on ecosystem function are beyond our full comprehension?

I suggest two criteria for successful copying. First, we should be especially careful when making changes that will be hard to reverse. This applies to human inventions as well as to ideas from nature. If we are considering a new herbicide to kill thistles in agricultural fields, does that herbicide break down quickly in soil? If not, what happens if we discover problems only after it has been used for some years? Even if we stop using the herbicide, it could persist in the soil, continuing to cause difficulties for many years.

For example, some squash grown on my brother Tom's organic farm was found to have small traces of dieldrin, decades after this insecticide was used by anyone for the last time, simply because dieldrin breaks down so slowly in the soil. Concentrations of dieldrin in the squash were extremely low. I estimated that someone would have to eat 500 pounds of squash at one meal to get a dangerous dose—imagine taking 500 times the recommended dose of aspirin!—but it's still worrisome that it lasted so long in the soil. The faster a pesticide breaks down, the less we have to worry about long-term effects.

Similarly, *biological control* using predatory insects can be a great way to control pests, but can we change our mind if something goes wrong? A weevil that was released to eat weedy thistles in farm fields developed a taste for some rare wild thistles that are important to butterflies.[135] How do we get rid of a "living pesticide" that actually reproduces rather than breaking down over time?

Second, we should ask whether the aspect of nature we are planning to copy has been reliably improved by natural selection or some other natural process. If so, then it may indeed be better, at least by nature's criteria, than anything we could design. If not, then blindly copying that aspect of nature on our farms does seem like cargo-cult agroecology.

## Learning from Leaf Blight

Since 1970 or so, most evolutionary biologists have agreed that the improving power of natural selection decreases as we move from genes and enzymes, to individual adaptations of plants or animals, to entire ecosystems.[136–141] This is why we should expect individual adaptations to be superior, on average, to ecosystem-level organization. (For the purposes of this book, I need to make this claim only for improvement by agricultural criteria.) Levels of organization between individuals and ecosystems, such as groups of plants or ant colonies, raise some interesting questions, which will be discussed in later chapters.

The decreasing power of natural selection as we move from genes to ecosystems has been explained most clearly by Richard Dawkins, in *The Selfish Gene*[142] and other books. Alleles competing for places in the next generation are typically found in different individuals of the same species, so natural selection tends to improve the survival and reproduction of individuals. Similarly, improvements in individual fitness may improve the productivity or sustainability of the ecosystems where they live, but this is not always the case. Nor is it always the case that the most successful alleles (those whose frequency increases most) always benefit individuals hosting those alleles. An interesting example comes from Southern corn leaf blight.

In 1970, much of the U.S. corn crop was destroyed by this fungal disease. This disaster is often used, rightly, to show the need for more genetic diversity within crop species. Having more genetic variability among corn varieties would have decreased the chance of them all being destroyed by the same disease.

The question of how to use crop diversity most effectively is more complex than it may seem, however. Which is more useful, diversity within fields, or diversity among fields? This question is deferred to later chapters, because the value of diversity is not the only lesson from the Southern corn leaf blight epidemic.

First, this epidemic was an evolutionary problem. Old genotypes of the fungus *Bipolaris* (previously called *Helminthisporium maydis*) caused little disease, even in the corn varieties devastated in the 1970 epidemic. A mutant fungus that attacked these varieties was apparently present in the United States for at least a few years, but natural selection did not particularly favor this genotype until the right combination of weather and susceptible host plants occurred around 1969.[143] Then the mutant strain increased rapidly in frequency, with devastating results.

Second, mistakes made by the seed industry contributed to the problem, but the industry also mobilized quickly to find solutions. They quickly grew corn for seed in the southern hemisphere, during the northern hemisphere winter, speeding the availability of disease-resistant varieties. The situation would have been much worse if U.S. farmers had had to wait for seed multiplication in the United States.

And last, this disaster illustrates how natural selection among plant alleles can act against the best interests of plants having those alleles. This is a somewhat complex story. First, you need to understand why corn plants all across the United States were similar enough, genetically, that they were all susceptible to the same fungus.

This genetic uniformity was an unfortunate side-effect of what had seemed like a major breakthrough in corn seed production. When a corn plant pollinates itself, the resulting seed is less vigorous than hybrid seed, which is produced by crossing two genetically different parents. To get this vigorous hybrid seed, companies used to grow two different genotypes together and let them cross. To prevent self-pollination, they had to remove the male, pollen-producing flowers (tassels) from one genotype. Seed harvested from that genotype was therefore a hybrid between that maternal plant and the plant that had pollinated it.

Removing tassels is a lot of work. (Tassel removal once provided a summer job for Bob Loomis, later a leading crop ecologist and one of my mentors. I wonder if that inspired his later work showing that shading by tassels can decrease photosynthesis.[83]) So when someone discovered a corn plant in Texas that made no pollen, its value was quickly recognized. Soon, the descendants of this plant were being used as substitutes for detasseled plants in hybrid seed production. This innovation lowered the cost of hybrid seed, benefiting seed companies and farmers. However, it also meant that most corn plants in the United States were descended, in the female line, from the same male-sterile ancestor. This genetic uniformity is why most U.S. corn plants were susceptible to the same fungus. Today, mechanical detasseling is again fairly common.

But this is also a story about selfish genes. In what way were some corn genes selfish, and why? Male sterility has been useful to seed companies, but is there any benefit to the plant? If we use an evolutionary definition of *benefit*, we can rephrase this question: do male-sterile plants have more descendants than plants that produce both pollen and seeds? Probably not. Instead, the chief beneficiary of male sterility appears to be the male-sterility allele itself.

How might an allele for male sterility persist over the course of evolution? In terms of whole-plant fitness (number of descendants, relative to competitors), the cost of pollen production is well worth it, given the opportunity to fertilize another plant. Depending on how much competition there is for pollination opportunities, a plant might actually have more descendants if it put most of its resources into making pollen grains, which are tiny and therefore cheap, letting other plants make expensive seeds. Under drought, some corn plants selfishly increase pollen production at the expense of seed production.[113]

Fitness looks different from the viewpoint of the male-sterility allele itself, however. The male-sterility gene is inherited only in seeds and not, of course, via pollen. Although a pollen grain is much cheaper than a seed, any investment in pollen still causes some decrease in resource availability for seeds. So if a gene that is transmitted only in seeds mutates into an allele that somehow reduces pollen production, that allele will slightly increase its own chances of making it into the next generation, at the expense of the plant's overall fitness.[144]

Which genes are transmitted only in seeds? Most plant genes are encoded in the DNA of the cell nucleus and transmitted in both seeds and pollen, but there are also a few genes in mitochondria. Because mitochondria are inherited only from the female parent, an allele in mitochondria that reduces or eliminates pollen production won't thereby decrease its own transmission. In fact, by freeing resources for seed production, it will somewhat increase its own transmission. So evolutionary biologists were not surprised when the gene for male sterility turned out to be in the mitochondria, not in the nucleus.

Nuclear genes have different interests, however, a point also demonstrated by this hybrid-corn example. In hybrid-seed production fields, the male-sterile plants were pollinated by the male-fertile genotype. But farmers typically grow only one genotype in a field, and that genotype would have to produce pollen; otherwise, no seed would be produced. So farmers couldn't grow a male-sterile (no-pollen) genotype.

Plant breeders solved this problem; they found a gene that restored pollen production. Seeds produced on a male-sterile maternal plant, pollinated in a seed-company field by a male-fertile plant with the restorer gene, resulted in vigorous plants that produced seeds in farm fields by self-pollination. And where was the restorer gene? In the nucleus.

Nuclear genes are transmitted in both seeds and pollen, so natural selection favors nuclear-gene alleles that promote whatever balance between seeds and pollen increases overall reproductive success. Therefore, the interests of nuclear genes rarely conflict with the interests of the whole plant. In fact, nuclear genes can act to restrain the selfishness of other genes, including those in mitochondria. The pollen-restorer gene is an example.

This seems to be the general pattern in all plants, not just corn: male-sterility alleles are found in mitochondria and transmitted only in seeds, whereas pollen-restorer alleles are found in the nucleus and transmitted in both seeds and pollen.[145] The Southern corn leaf blight epidemic was an indirect side-effect of conflict between nuclear genes and mitochondrial genes.

## Natural Selection Improves Genes and Individuals, Not Ecosystems

The male-sterility allele in corn mitochondria was "improved" by natural selection only in the sense that it improved its own transmittal to the next generation, via seeds alone, at the expense of nuclear genes transmitted in both pollen and seeds. Usually, however, natural selection favors whichever alleles are most beneficial to a plant's overall fitness. This is because, for most genes, the only way an allele can increase its representation in the next generation is to increase the survival and reproduction of plants in which it is found.[142] With few exceptions, therefore, we expect those alleles that are most beneficial to individual plants to be most common.

For example, most land plants have very similar versions of the photosynthetic enzyme rubisco. It is therefore reasonable to assume that no minor variation on that version would be consistently better for individual plants, at least in past environments. Minor variations must have arisen repeatedly over the course of evolution and competed with each other. Today's plants, including our crops, inherited their rubisco alleles from the winners.

Similarly, if a plant has thorns, it is safe to assume that its ancestors benefited from having thorns, because those ancestors must have outcompeted variants without thorns. The assumption here is that growing thorns requires an entire suite of genes. With so many mutation targets available, natural selection could easily eliminate thorns, if they were useless to the plant. Evolving thorns in the first place is presumably more complicated, however, so lack of thorns in other plants doesn't prove that they would not have benefited from thorns.

So natural selection among alleles usually benefits individuals. But just as natural selection favored the interests of male-sterility alleles over the interests of corn plants, natural selection favors individual fitness over ecosystem productivity, efficiency, or sustainability.

Often, improvements in individual fitness (more-efficient photosynthesis, for example) will also improve ecosystem function, just as more-efficient enzymes usually improve individual fitness. But individual selection can also have negative effects on ecosystems. Bark beetles that obligingly stay home, to prevent the spread of fungal disease to trees where they feed, would have few offspring. Natural selection can't look ahead to see long-term consequences for trees or beetles, any more than a river can flow uphill to find a shorter route to the sea.

To overcome negative effects of individual selection on ecosystems, when there are tradeoffs between individual fitness and ecosystem function, we would need some mechanism that reliably replaces poorly performing ecosystems with copies of better-performing ones. Individual selection is based on competition. (This includes indirect competition, like plants "competing" to be a little less attractive to pests than their neighbors.) But ecosystems are not analogous to organisms,[146] and they don't compete, directly or indirectly, with other ecosystems.

Some of the species in more-productive ecosystems may spread (at various rates) to less-productive ones, but spatial patterns responsible for differences in productivity are not replicated with anything like the accuracy of DNA-based inheritance of better-performing alleles. If a forest is destroyed by fire, because of the mix of tree species that grew there or their spatial patterns, the fire-resistant ecosystem next door will not move in *en masse*. Instead, the space will be colonized by species whose seeds are most easily transported by wind or birds, plus any seeds the "dead" ecosystem left behind in the soil. The most successful species among these colonists will be those that benefit most from newly open conditions, not those that are most fire-resistant.

Similarly, if wolves recklessly kill all the deer in their valley, they will not necessarily die out and be replaced by mountain lions that are better stewards. Instead, they will move into the next valley. An organism's genes are usually inherited together or not at all, but an ecosystem's species mostly migrate independently.

Ecosystem properties are, of course, affected by the genes of species that live there. For example, disease-resistance genes in trees may affect the growth and health of a forest. My focus, however, is on ecosystem-level properties that are not affected much by any single gene.

Consider a tropical forest with hundreds of tree species. Ecosystem-level properties include the number of species, the distribution of leaves with height or roots with depth, the fraction of total available sunlight absorbed by leaves, the number of months per year in which edible fruit is available, or the percent of rainfall that flows out of the forest in rivers. These ecosystem-level properties are influenced by millions of genes in thousands of species.

Some allele may increase the overall productivity of the forest slightly, but that will not increase the *relative* success of that allele in its indirect competition with alternative alleles in the same forest. The allele will increase in frequency only if it gives individual plants with that allele an advantage over members of the same species with alternative alleles. So we expect natural selection to have improved individual plants much more than it has improved ecosystem-level patterns.

If ecosystem-level properties have not been consistently improved by natural selection, what does that imply for our search for nature's wisdom? Suppose you are planting an orchard near a natural forest. A single species of tree (perhaps redwoods) dominates that forest. Should you assume that an orchard with only one species of fruit tree will be the most productive, efficient, and sustainable? No. Forests have not competed against other forests, based on productivity, efficiency, sustainability, or any other criterion. Furthermore, forests do not inherit a number-of-species gene, a leaf-distribution-with-height gene, or a ratio-of-predators-to-prey gene from successful forests of the past. Natural selection is good at making rubisco more efficient or thorns sharper, but it is powerless to adjust the number of tree species in a forest up or down for the benefit of the ecosystem as a whole. Therefore, you should base the number of tree species in your orchard on something more than simply copying the number of tree species in nearby forests.

A careful study of the redwood forest might show that the low tree diversity has some advantages, or it might not. Comparisons with other forests (and orchards!) would probably be worthwhile. Considering differences between your goals and those of natural selection is also important. These topics will be discussed in more detail in later chapters. For now, my main point is that you cannot simply assume that a natural forest has the optimum number of tree species for farms in your bioregion.

Typically, agroecologists may choose more-diverse ecosystems to copy. But if we assume that natural processes always drive ecosystems to whatever composition is optimum for local conditions, surely that argument would apply to ancient redwood forests as well. Redwoods are not the only example of low plant diversity in natural ecosystems. Wild relatives of important crops, like rice and wheat, grow in near monocultures.[147] This could be seen as evidence that these crops should be grown in monoculture, but only if we assume that the structure of natural ecosystems is optimal.

Remember the second of my three core principles: testing by mere persistence is weaker than competitive testing. So the persistence of monocultures in nature shows that monocultures can be sustainable. Similarly, the persistence of diverse tropical forests shows that diversity and sustainability can be compatible. But, in each case, persistence is not evidence of improvement over millennia or of optimality.

## Comparing Natural and Agricultural Ecosystems

The preceding analysis leads to the conclusion that natural selection has not consistently improved ecosystem-wide properties, at least by criteria relevant to agriculture. But perhaps we should look at a wider range of data. What if there are processes other than natural selection that have improved the organization of natural ecosystems, in ways that would make them good models for agriculture? To test this hypothesis, we would like to compare a large number of natural ecosystems with agricultural ecosystems designed by humans. A fair test is more difficult than it sounds, however.

First, we need to agree on the criteria by which ecosystems can be compared. We could certainly think of criteria that would favor the natural ecosystem. For example, where ecotourism is a major source of income, "looking like a rain forest" could be a reasonable goal, and one that even the most sustainable farm would be unable to meet.

Or maybe not. When I was about nine years old, my father took the family along on a field trip to Costa Rica. I was a big fan of Tarzan and looked forward to seeing the "jungle." Nothing we saw seemed to match what I'd seen in television, however. Finally, one day, I looked around and said "at last, a real jungle!" Dad laughed. "Look more carefully; we're on a farm." Sure enough, the vegetation was dominated by coffee and banana trees. When I lived in the coal country of West Virginia, a guest came back from a walk, enthusiastically describing a park-like area he'd discovered, back in the woods. It was a reclaimed strip mine, planted to grass. Aesthetic criteria can be tricky.

Instead, I suggest that we compare ecosystems using criteria from chapter 2. Which can export more food to feed urban populations: natural ecosystems or human-managed ones? Which can do so most consistently (that is, with least variation among years) and sustainably (with no downward trend over decades)? And what about environmental impacts, like water pollution? I wasn't able to find all the data needed to answer these questions, but what I did find made me doubt that either known or unknown mechanisms have optimized the organization of natural ecosystems to maximize productivity or stability.

Agricultural ecosystems can certainly be more productive than natural ones, but this doesn't necessarily show that their ecosystem-level organization is superior. Agricultural ecosystems typically receive nutrients (fertilizers) or water (irrigation) not supplied to their natural counterparts. Even without extra inputs, their greater productivity may come from crop genotypes with greater allocation to grain, rather than more sophisticated deployment of species in space or time. So, ideally, we would like to compare the productivity of wild species managed by humans without artificial inputs, relative to the productivity of the same species in natural ecosystems. Wild rice and reindeer/caribou comparisons seem to come closest to meeting these criteria, but I will discuss some less-perfect comparisons first. Decades ago, at the Cedar Creek Natural History Area, where some of my colleagues now study biodiversity, researchers compared the productivity of prairie and forest plots to a nearby corn field.[148] The corn field received nitrogen fertilizer, while the prairie relied mainly on symbiotic nitrogen fixation by wild legumes. Was the organization of the natural prairie (including the ratio of nitrogen-fixing legumes to nonlegumes) sufficiently superior to overcome benefits from fertilizer?

Overall growth of the corn plot was 9456 kilograms per hectare (kg/ha; note that a kg/ha is about one pound per acre) per year, versus only 930 kg/ha for the prairie plot. One implication of this tenfold difference is that, if we ever develop economical methods to convert whole plants to ethanol, we would need to harvest 10 acres of unfertilized prairie to get as much plant material as from 1 acre of fertilized corn. The *harvest index* (ratio of grain to total aboveground biomass) for corn is about 50 percent, so even if we harvested the corn grain for food and converted only leaves and stalks to ethanol, we would still get about five times as much material as from the same area of unfertilized prairie.

Differences in seed production were even more dramatic: 5337 kg/ha for corn versus < 1 kg/ha for prairie plants! The nearby oak forest did a little better, producing 20 kg/ha of acorns.[148] If our goal is grain production, these data suggest that we could get the same amount of grain from one acre of fertilized corn as from thousands of acres of unfertilized prairie plants. With nitrogen fertilizer, an acre of corn presumably causes more water pollution (specifically nitrate) than an acre of prairie. But it seems unlikely that it causes more nitrate pollution than the thousands of acres of natural prairie that would apparently be needed to produce the same amount of seed protein.

What about consistency over years? Another study of native prairie found enormous variation in seed production, ranging from 7 to 525 kg/ha over a 3-year study.[149] Even the worst year was better than reported earlier for Cedar Creek, with its nutrient-poor sandy soils. But seed production in this prairie, even in the best year, was a fraction of the average 1988 yield for grain crops worldwide, which was 2480 kg/ha.[11]

In the Long-Term Research on Agricultural Systems (LTRAS) experiment, our wheat yields from 1996 to 2002 averaged 4069 kg/ha without irrigation or fertilizer (trending down as soil nutrients were used up) and 5779 kg/ha with both inputs.[30] Organic and conventional corn yielded even better, at 7559 and 11,484 kg/ha, respectively, albeit with major inputs of irrigation water and either composted manure or synthetic fertilizer.

But prairies are presumably sustainable, in their natural state. What about agriculture? At LTRAS, conventional corn showed a significant positive yield trend over years, suggesting that this system (corn alternating with tomato) may be sustainable as well as productive. I would like to see a few more decades of data before relying on this conclusion, however, given the reversal of yield trends seen at Rothamsted,[47] discussed in chapter 2. Similarly, I would like to see how prairie productivity might change, if harvested material were removed every year. In other words, prairies may be sustainable if left alone, but can they be harvested sustainably for biofuels or other uses?

Other comparisons also show greater productivity for agricultural ecosystems, relative to natural ones. It has been estimated, for example, that tropical forests have only 1 to 2 kg/ha of harvestable tubers,[150] whereas a farm growing cassava can yield 12,000 to 29,000 kg roots/ha (fresh weight).[151] Oak forests produce only 0.4 to 55 kg/ha of acorns,[152] whereas almond orchards can produce 926 to 1746 kg/ha of nuts.[153] Note that the almond orchard yield was not only greater, but also more consistent than the acorn yield. Would we have enough self-discipline to limit our population to what could be supported by 0.4 kilogram of acorns per hectare? If not, would we have enough foresight to store enough acorns during good years to eat in bad years? Almonds also taste better than acorns, but that can't be attributed to ecosystem-level organization.

If the results described in the preceding few paragraphs are typical, then it is clear that natural ecosystems are less productive, whether we consider total biomass, seeds, tubers, or nuts, than well-managed agricultural ecosystems. However, these data do not necessarily prove that agricultural ecosystems are better organized. Most of the agricultural fields (except the nonirrigated, unfertilized wheat control plots at LTRAS, in which yields decreased over years, showing a lack of sustainability) received inputs of water and nutrients not supplied to the natural ecosystems.

Furthermore, greater seed production, relative to natural ecosystems, could be the result of crop genotypes that put a larger fraction of their resources into seeds, relative to wild plants. Although this would not explain the greater biomass production of corn relative to prairie, the greater seed production of agricultural ecosystems could have little to do with human-imposed arrangements of plants in space or time.

In noting the limitations of these comparisons, I am not arguing against the use of inputs or high-yielding crops, just pointing out that these variables make it difficult to compare the ecosystem-design skills of humans with whatever processes structure natural ecosystems. We need better comparisons.

## Fairer Comparisons of Ecosystem Design: Wild Rice

An ideal comparison of the processes that organize natural ecosystems versus the agricultural-ecosystem design skills of humans would require some constraints on inputs. For research purposes, at least, we would want to forgo fertilizer and irrigation. Also, we would want humans to limit their selection of plants and animals to the same wild species found in whatever natural ecosystem is used for comparison.

Given these two constraints, can we harvest more food, sustainably, from human-managed ecosystems or from natural ones? I identified two systems that could, in theory, meet both criteria: human cultivation of wild rice without fertilizer, and human management of reindeer herds (domesticated caribou), discussed in the next section. The ideal controlled experimental comparisons do not appear to have been made, however.

Wild rice, *Zizania aquatica*, is not closely related to Asian rice. It grows naturally as a near-monoculture,[154] in lakes from Minnesota to Canada. Wild stands have been harvested for hundreds of years, so this traditional practice appears to be sustainable. Hand-harvested yields of wild stands tend to cycle, however, with each high-yield year followed by a "crop failure" and then two or three recovery years, so natural wild rice scores poorly on our stability scale.

In years when there is enough grain to make harvesting worthwhile, yields of natural wild rice stands in Minnesota average 30 to 40 kg/ha.[155] Yields of human-managed wild rice in Minnesota are often much higher than wild stands: up to 1800 kg/ha,[156] with less variability. Some of this difference is due to addition of nitrogen fertilizer, however, albeit in small amounts relative to most crops. Management of wild rice by humans also includes some innovative practices not seen in nature, such as plowing insulating snow from ice-covered lakes to kill perennial weeds by freezing. Sometimes, humans also control water depth.

Furthermore, humans mostly plant wild rice genotypes that hold onto their seeds, rather than dropping them, which presumably increases the fraction of seeds that gets harvested. This genetic difference may also contribute to the faster cooking and reportedly better flavor of wild-harvested wild rice.[157]

The only data I found on yields of unfertilized wild rice managed by humans is unpublished: a seed yield of 100 kg/ha, sent to me by Dan Marcum, a farm advisor associated with the University of California. Although this is more than double a good-year yield for wild stands in Minnesota, differences in climate could be responsible. Data for human-managed, but unfertilized, wild rice in Minnesota are probably out there somewhere.

Much of the yield difference between farmed wild rice and natural stands is clearly due to nitrogen fertilizer. On the other hand, none of the data I found on wild rice indicated that natural processes "organize ecosystems for the maintenance of high productivity and diversity," as has sometimes been proposed.[158] Wild rice stands are less productive than human-managed wild rice and less reliable (one good year in four). For what it's worth, they have similarly low plant diversity.

## Human-managed Reindeer versus Wild Caribou

What about systems based on grazing animals? Many such systems use animals that have been substantially modified from their wild ancestors by selective breeding. "Surreptitious mating between aurochs bulls and domestic cows," introducing Y chromosomes still detectable today,[159] would have slowed evolutionary divergence between domesticated and wild cattle, but only until the wild aurochs went extinct. Significant external inputs are also common in grazing systems, as when forage is supplemented with purchased grain.

Herds of reindeer managed by humans may come closest to meeting the criteria needed to compare human ecosystem-organizing skills with whatever processes organize natural ecosystems. Reindeer are domesticated or semidomesticated caribou. Today, most reindeer receive supplemental feed.[160] But traditional reindeer herds ate the same food as caribou, mostly lichens. Older data are therefore more useful in comparing the management skills of humans with the processes that structure natural ecosystems.

Even without supplemental feed, human-managed reindeer herds appear to have been a more productive source for meat for human consumption, relative to hunting of wild caribou. I am not taking a position on the morality of killing animals for meat, just trying to test the hypothesis that natural processes organize ecosystems better than humans can, particularly by criteria relevant to agriculture. As always, to "test" a hypothesis means to subject it to the possibility of disproof.[64]

The density of wild caribou in Quebec has reportedly never exceeded 1.5 animals per square kilometer.[161] The maximum rate of caribou harvest by hunting in recent decades was 7 percent per year,[162] so less than $1.5 \times 0.07 = 0.1$ animal was harvested per square kilometer per year. Herded reindeer in Finland have almost always had densities greater than 2 animals per square kilometer.[160] Based on annual harvest rates of 25 to 35 percent for adults and 40 percent for calves,[163] production cannot be less than 0.5 adult per square kilometer per year, plus calves. So production of meat (not counting what is eaten by wolves) is more than five times as great for herded reindeer as for wild caribou.

But has the greater productivity of human reindeer management come at the expense of more year-to-year variability or less long-term sustainability? For wild caribou, historical data show that large fluctuations in herd sizes have occurred at multiple locations, over periods of 80 to 100 years.[161] Wild caribou populations often overshoot sustainable levels (the *carrying capacity*), then crash.

When there are too many caribou, they overgraze the lichens. The resulting food shortage leads to reduced survival and also reduced pregnancy rates from poor nutrition, bringing population size back down again. But these ecosystem-level negative feedbacks haven't been optimized by competition among ecosystems with different feedback loops. This contrasts with how natural selection has optimized negative feedbacks that control our heart rates, which have been improved by direct or indirect competition among individuals. So ecosystem-level feedbacks are too weak, or too slow, to prevent population crashes.

A good patch of lichens can represent up to several decades of growth. What happens when the caribou population gets big enough that it takes more than one year's new lichen growth to feed them all? The situation is unsustainable long-term, just as harvesting trees faster than they grow back is unsustainable long-term. But in the short term, caribou survive and reproduce, eating lichens that grew over a period of many years, perhaps ignoring warnings from "peak lichen" alarmists, just as we ignore warnings about "peak oil" or (even more worrying) "peak phosphorus." Only when the lichens are gone does the caribou population crash. For example, the size of one wild caribou herd in Quebec increased from less than 200,000 in the late 1970s to a peak of about 750,000 animals around 1988, then decreased to about 450,000 in 2001 (60 percent of peak), with some observations suggesting that lichens were starting to recover.[164]

If wolves and other natural feedbacks fail to prevent caribou population fluctuations and crashes, can humans do better? In traditional reindeer herding, oscillations in herd size were also common, but somewhat less severe than in nature. Herds in Finland increased from 120,000 to 286,000 and then decreased to 215,000 (75 percent of peak) from the 1970s through the 1990s.[160] Once again, overgrazing of natural lichen pastures was a major cause. Reindeer researcher Jouki Kumpula estimated that to have enough lichens for maximum reindeer growth, lichens would first need to be left ungrazed for 18 years.[165]

Overgrazing by human-managed herds has apparently not caused a collapse in meat production, however. Recently, herders have increased their use of supplemental feeding, an external input that reduces our ability to compare reindeer managed by humans with their wild caribou counterparts. But increased slaughtering of calves for meat has also helped stabilize populations.[166] Reindeer without calves don't have to feed them or share forage with them, so they are more likely to survive the winter.

Herders also manage reindeer in creative ways that nature can't. Consider the ratio of females to males. A mostly female herd, with only enough males to fertilize them all, would have double the potential growth rate. This is what John Maynard Smith called "the twofold advantage of not producing males."[167] A mostly female caribou herd would convert lichen to calves for meat more efficiently, because most adults would be producing calves, rather than fighting over females. Natural selection, however, has given caribou genetic mechanisms that produce roughly equal numbers of males and females. The same is true for many other species, with some interesting exceptions.[168]

A selfish-gene perspective explains why sex ratios are usually balanced. In a mostly female population, the average male will have more descendants than the average female. Therefore, any mutant allele that biases offspring sex ratio to produce more males would quickly spread, until a 50:50 ratio was restored.[142] But caribou herders aren't always bound by natural selection's constraints. They preferentially slaughter male calves, giving the adult herd the optimum, female-biased sex ratio. Using this and similar tricks, reindeer herders have produced more meat and kept populations more stable, relative to wild caribou herds, even without supplemental inputs.

Results for both wild rice and caribou/reindeer suggest that human-managed ecosystems can meet key agricultural goals (productivity, sustainability, stability) at least as well as natural ecosystems. So, before agriculture copies landscape-scale patterns found in natural ecosystems, we would need better evidence that specific patterns will help us reach our agricultural goals.

## Perspective

Any complex system will have interactions among its parts, resulting in various kinds of feedback. It would not be surprising to find some beneficial negative feedbacks in many natural ecosystems, conferring some degree of stability on one or more ecosystem properties. But these feedbacks are not analogous to the feedbacks that control our body temperature or blood sugar. Throughout our evolutionary past, individuals whose bodies had slightly different feedback control systems (based on different combinations of alleles) competed against each other, directly or indirectly. Those with the best regulation (ensuring stability when needed, but also responding appropriately to danger) tended to win those competitions. Ecosystems have not been improved in this way, because they have never competed against other ecosystems for the chance to reproduce. Furthermore, the "inheritance" of landscape-scale feedbacks is much less reliable than DNA-based inheritance.

But have mechanisms other than natural selection consistently improved patterns or feedback mechanisms in natural ecosystems, making them useful models for agricultural ecosystems? The available data (orchards versus forests, wild rice, reindeer/caribou) suggest that the organization of natural ecosystems is not consistently better than the organization of agricultural ecosystems. But those data aren't perfect. Additional research might be worthwhile, such as a side-by-side comparison of the same unfertilized wild rice genotype with and without human management. Based on the evolutionary theory and field data presented in this chapter, however, it seems unlikely that blindly copying the organization of natural ecosystems would yield optimal designs for agricultural ecosystems.

# 7

## What Won't Work

MISGUIDED MIMICRY OF NATURAL ECOSYSTEMS

IN THIS CHAPTER, I will discuss four of the most-popular ideas for improving agriculture that have been inspired by nature: perennial grain crops, reliance on only local sources of nutrients, *polyculture* or *intercropping* (that is, deploying crop diversity as mixtures, as in many natural ecosystems), and reliance on biodiversity to control pests. All of these were suggested in one classic paper by Wes Jackson and Jon Piper, "The necessary marriage between ecology and agriculture."[14] I will argue that their ideas are representative of many self-styled "agroecologists." I will then discuss each of their proposals in light of my conclusion from the previous chapter—namely, that copying landscape-scale patterns from natural ecosystems is not necessarily a good idea. I conclude that while each of these ideas has potential, other approaches may work as well or better.

## Applying Ecological Principles to Agriculture

When I was at UC Davis, one seminar speaker began by asking us for "ecological principles" that could be applied to agriculture. Although the audience included many people with expertise in both ecology and agriculture, we were pretty slow in coming up with ecological principles that are as fundamental as those from other scientific disciplines.

Fundamental principles of physics include conservation of matter and energy. Chemistry has the periodic table, based on atoms as the building blocks of matter. Biology has the central dogma of molecular biology (DNA sequence controls amino acid sequence in proteins, never the other way around), cells as building blocks for complex organisms, and evolution by natural selection. Ecology inherits all of these principles from chemistry[169] and other disciplines, of course, but surely there must be additional principles specific to ecology. For example, engineering inherits principles from physics, but adds beam theory, which predicts how bridges bend under load. Medicine inherits principles from biology, but adds the germ theory of infectious disease. What about ecology?

After some hesitation, people in the audience called out things like "competition!" and "food webs!" These are topics that ecologists have done much to explain, but they aren't exactly *principles*, comparable to "infectious disease is caused by microorganisms, not witches," for example. So what are the central principles of ecology? The latest edition of a leading ecology text provides a clue. It begins with a chapter titled "The Evolutionary Backdrop."[170] Evolutionary principles are indeed fundamental to ecology.

Given a little more time, I would have suggested "selfish genes" as a key evolutionary principle inherited by ecology: the alleles that spread are those that facilitate their own increase or reduce the success of competing alleles, regardless of effects on ecosystems. But I'm not sure the speaker would have agreed. Instead, her proposed principles were very similar to those given by Jackson and Piper.[14]

Jackson and Piper began by proposing that agriculture should rely less on "human cleverness" and more on "nature's wisdom." I suggest that we need more of both, especially if human cleverness is tempered by human wisdom. But where is nature's wisdom to be found? As discussed earlier, I would look first to the sophisticated, competitively tested adaptations of individual wild plants and animals.

They proposed, however, that the "relics of pre-Columbian [not pre-Indian?] vegetation that remain must serve as our best standards" and that "patterns and processes discernible in natural ecosystems still remain the most appropriate standard." If *vegetation* means plant communities, rather than individual plants, and if patterns and processes in natural ecosystems refers to ecosystem-level processes—nitrogen movement through a watershed, for example—then Jackson and Piper are proposing entire natural ecosystems, not the individual adaptations of wild plants, as standards for agriculture.

But, as I argued in the previous chapter, the overall organization of natural ecosystems may be an inadequate standard. Ancient ecosystems may have met our minimal standard of sustainability, but there is no reason to expect them to be organized for optimal productivity, reliability over years, or efficiency in the use of scarce resources.

I will focus this chapter on the specific proposals and wording from Jackson and Piper[14]—perennial grain crops, local sources of nutrients, intercropping, and reliance on biodiversity to control pests—because these ideas are quite popular among agroecologists. There have been hundreds of studies of intercropping, for example,[171,172] so I think it's reasonable to take this paper as representative.

## Would Perennial Grain Crops Be Better?

Despite the preceding references to vegetation and ecosystems, Jackson's and Piper's first specific suggestion focuses on wild plants rather than ecosystem structure. This is superficially consistent with my own focus on natural selection's competitively tested individual adaptations of wild species.

Specifically, they advocate "development of perennial grain crops," a goal that has been the main focus of Jackson's Land Institute for many years. This wouldn't be high on my list of promising approaches, as I will explain, but it does build on past natural selection in ways that copying entire natural ecosystems (a prairie, say) would not. If the native vegetation of a given region is dominated by perennials, that is reasonable evidence that they had some advantage over annuals. By analogy with my previous arguments, more-perennial plants must have competed repeatedly against less-perennial mutants, and won. As always, however, natural selection's winners aren't necessarily better suited to our agricultural needs.

Some of the ways in which perennials may have a competitive advantage, however, do seem relevant to agriculture. Consider early spring growth. Annual plants, growing from seed, may not have enough leaf area to capture of all the available sunlight, at least at first. This is because a seedling that depends on only its own photosynthesis for growth will produce new leaf area relatively slowly. Perennials, on the other hand, may use stored photosynthate from last year to quickly grow more leaves, capturing most of the available sunlight earlier in the season.

For example, a perennial grass called *Miscanthus* can reportedly produce more total biomass than corn, an annual grass.[173] These two species have similar efficiency in converting sunlight to biomass, but the perennial captures more sunlight early in the season, perhaps using leaves whose growth was subsidized by photosynthate stored the previous year. Interestingly, *Miscanthus* also captures more sunlight later in autumn, so differences other than perenniality (for example, greater cold tolerance) may also be important.

Lee DeHaan, of the Land Institute, has pointed out that perennials may also have access to more water than annuals, especially on sloping land. Perennials often have deeper roots, and they may slow the downhill movement of water over the surface better than annuals do, allowing more of it to soak in. If, in a given region, annuals have competed against perennials repeatedly and lost, maybe it's because the perennials were able to use available resources more completely. Whatever the explanation, we should at least consider increased agricultural use of perennial crops in that region.

This same logic, however, makes me question the Land Institute's emphasis on increasing *seed* production from perennials, to convert them to grain crops. Converting perennials to annuals has sometimes been successful (with cotton, for example), but it's worth asking why natural selection in prairie perennials has favored roots and stems over seeds.

Jackson and Piper praise "evolutionary theories on life history strategy and resource allocation." Those theories[90, 91] predict tradeoffs that make it impossible to maximize both reproduction (for example, grain yield) and longevity (key to the benefits of perenniality). Tradeoffs based on conservation of matter make the same prediction. Sugars allocated to seeds cannot be used to keep the plant alive during periods when freezing weather or drought limit photosynthesis.

The longevity-versus-reproduction tradeoffs predicted by evolutionary theory and conservation of matter can sometimes be difficult to detect in the field. Some tradeoffs are apparently less severe than expected. Seed production sometimes triggers increases in photosynthesis (perhaps by opening stomata wider, recklessly transpiring more water?), making more photosynthate available both for immediate seed production *and* long-term survival. Seed production can also *decrease* photosynthesis, however, by competing with rubisco for nitrogen.[174]

Some seed-containing fruits photosynthesize, meeting some of their own photosynthate needs, but this isn't a big advantage if they shade leaves that would have had higher rates of photosynthesis. Individual plants that are larger because they happened to grow in a better spot can allocate more resources to both seeds and future survival. This can create a misleading positive correlation, overall, between seed production per plant and growth or survival, even though the options available to each individual plant involve a negative relationship between reproduction and survival.[175, 176]

Despite these complications, a recent review found that "case studies . . . generally support the predictions of the cost of reproduction hypothesis."[177] Ragweed plants that made fewer seeds were much more likely to survive than plants of the same size that made more seeds.[178] Removing flowers from one wild perennial significantly increased its survival rate,[179] consistent with a tradeoff between seed production and survival. In two other wild species, there was no statistically significant effect of flower removal on survival, but this may have been because survival rates were so high. Only 4 of 349 plants died: one with flowers removed and three allowed to make seeds.[180]

I have turned annual kale into a "perennial," harvesting leaves from the same plants for several years, by removing and eating the flowers. The tops are killed by freezing weather every winter, but new sprouts from the roots give us fresh leaves long before seed-grown kale would be ready to harvest.

Ecologist Jose Obeso suggested that "perennial plants should maintain reproduction at low levels, which has minor effects on growth and survival."[177] A perennial plant could produce more seeds in a given year by using more of its stored resources—putting its long-term survival at risk—only to have all of its seedlings destroyed by drought, fire, or grazing animals. It may be safer to produce a few seeds each year, over many years. That's what many perennials do, as shown by the pathetic seed production (but good survival) of oak trees and many prairie plants, discussed in the preceding chapter.

But is some increase in seed production possible, without losing perenniality? Starting from a low baseline, a small increase might increase the risk of dying only slightly. For example, a perennial that puts only 5 percent of its photosynthate into reproduction might be able to double its seed production to use 10 percent of photosynthate, with only a small increase in the risk of early death. Plant breeders working to raise seed yields of these plants might therefore achieve significant increases at first—a doubling in seed yield, wow!—unknowingly accepting small increases in risk in exchange for increased yield. For seed yields to approach those of annuals, however, plants would have to put all available resources into seed production at the end of the growing season, as annuals do, making death almost certain.

So much for theory. How much actual progress has been made in breeding perennial grains? A previous 25-year effort to develop perennial wheat resulted in varieties with grain yield comparable to the lowest-yielding annual wheat varieties in California, but only half the yield of a reference variety in Oregon, according to a 1963 report.[181] Annual wheat yields have more than doubled since then, so a generous estimate would put the yield of this perennial wheat at 25 to 50 percent that of current annual wheat yields.

Potential advantages of a perennial crop include saving the cost of planting seed each year. However, "with only 30 percent of the stand surviving in the third year, the Oregon crop did not compete well with cheatgrass."[181] The need for frequent reseeding of a somewhat-perennial crop would reduce its potential cost savings. Seeding also disturbs the soil, contributing to erosion. Low-till seeding methods can reduce this disturbance, but those methods are already used with annual crops.

A recent paper from researchers at the Land Institute promotes the prairie legume bundleflower as a perennial with seed yield of "up to" 1700 kilograms per hectare (kg/ha) in Kansas.[182] This species was also studied here at the University of Minnesota. The average seed yield here was less than 500 kg/ha (55 grams per plant, with 0.76 × 1.52 meter spacing), even with herbicides and irrigation,[183] consistent with some other yield reports for potential perennial grain crops.[184] This is only 5 percent of Minnesota's current average corn yield of about 10,000 kg/ha. So it would take 20 times as much land to grow the same amount of grain.

Furthermore, survival was only about 50 percent, even for plants that were never harvested. So bundleflower, like perennial wheat, has both much lower grain yield than existing annual crops, and poor enough survival that it would require frequent reseeding. With enough breeding effort, it might be possible to bring the seed yield of bundleflower up to that of soybean, but only by turning it into an annual, like cultivated soybean. The breeder involved, Lee DeHaan, has moved on from bundleflower to perennial wheat. Initial progress has been promising, consistent with my suggestion earlier that it's much easier to double a low seed-production rate than it will be to approach the yields of annual crops.

Perennials certainly have potential. But given the tradeoff between perenniality and seed production, emphasis on grain production may be misplaced. Wild perennials are really good at surviving and producing leaves and stems, rather than seeds. That's what natural selection seems to have chosen. Can we find a use for leaves and stems?

We could let cattle, sheep, or bison eat the leaves and stems as forage, then eat the animals or drink their milk. Although these animals convert only a fraction of the forage they eat to meat or milk, they can digest a much larger fraction of a plant than we can. In Minnesota, some of the same researchers who found bundleflower *seed* yields of less than 500 kg/ha got seasonal *forage* yields up to 5000 kg/ha.[185] Based on these numbers, we would need only 10 percent conversion efficiency of bundleflower forage by animals to get as much food as milk or meat as we would from eating bundleflower seeds directly. Dairy cows can be about 23 percent efficient in converting forage calories to milk calories, and 29 percent efficient in protein conversion.[16] So we might be able to get three times as much protein, as milk, from animals eating the whole plant, as we could from eating the grain ourselves. Incidentally, it would be a mistake to assume that a new grain crop would necessarily be eaten by humans rather than animals; bundleflower seed has already been tested as feed for chickens raised for meat.[186] Only some human populations, descended from herders in Europe or Africa,[187, 188] have evolved the ability to digest milk as adults. Some people have religious or ethical objections to eating meat. But assuming that some people will continue to drink milk or eat meat, would bison grazing native prairie species be an interesting alternative to cattle eating corn (or bundleflower) grain in feedlots?

Herds of bison grazing on mixtures of native plants would resemble the original prairie ecosystem much more than a somewhat-perennial grain farm would. As discussed in the previous chapter, we should not expect this resemblance to natural prairies to guarantee greater productivity or sustainability. However, if restoring prairies has intrinsic aesthetic, cultural, or scientific value, then perhaps bison herds could play a part.

## Should Agriculture Use Only Locally Derived Nutrients?

Jackson's and Piper's other ideas are more closely tied to their apparent assumption that some natural process has optimized the organization of entire ecosystems, not just the traits of wild species. For example, they suggested that "truly sustainable agroecosystems . . . [are] those relying primarily on sunlight and locally derived nutrients."[14] This seems to conflate sustainability (undermined by reliance on *nonrenewable resources*, local or otherwise) with local autonomy, although the energy cost of transporting nutrients is a legitimate concern.

Natural ecosystems usually have no alternative to locally derived nutrients, but should agriculture copy this limitation? Most of the energy used on farms already comes from the sun and is captured via photosynthesis. However, most farms (with the possible exception of grazing-based animal agriculture) also use significant amounts of energy to power tractors or irrigation pumps. Long-term reliance on fossil fuels is clearly unsustainable, whether we are talking about the 5 percent used by farms and fertilizer factories[16] or the 95 percent used by the rest of the economy. sustainable options could include wind turbines and tractor fuels derived from crops.

Increasing reliance on locally derived nutrients would be more of a challenge than self-sufficiency for energy. As discussed in chapter 2, transfer of nutrients via farm products sent to cities (or to where meat, milk, or egg production is concentrated) tends to cause nutrient shortages in crop-growing areas. At the same time, these transfers cause surpluses in cities and livestock-producing areas, sometimes leading to water pollution, as excess nitrogen and phosphorus ends up in rivers or groundwater.

Even if transporting manure had no energy or economic cost, there isn't enough of it to meet all the nutrient needs of our crops. For example, the total animal manure supply in California is about one ton per irrigated acre,[189] a fraction of the several tons per acre typically applied by organic farmers. In the United States, the total supply of nutrients in sewage (which may contain pathogens and drug residues) is only about 2 percent of that in animal manure.[190]

Although meeting 100 percent of our crops' nutrient needs from local sources may prove to be impossible, there are some options worth exploring. These partial solutions differ somewhat for nitrogen, phosphorus, and other nutrients.

Nitrogen, the nutrient most commonly applied as fertilizer, is used by crops in large amounts, particularly to make protein. Local sourcing of nitrogen, as proposed by Jackson and Piper, may be possible. For example, the inexhaustible supply of nitrogen in the atmosphere could be converted to forms usable by crops by small-scale local chemical plants powered by windmills.[52]

A natural version of this process is carried out by symbiotic rhizobia bacteria in root nodules of legume crops like soybean and alfalfa. It may be possible to improve the efficiency of this process significantly,[29, 191] as discussed in chapter 9. However, complete reliance on biological nitrogen fixation would require much more widespread use of legumes in agriculture than we have today. This in turn would require major dietary changes in many countries: more beans, lentils, or tofu, or more meat or milk from animals eating clover or alfalfa. The issue isn't whether you, the individual reader, would willingly adopt such a diet, but whether hundreds of millions of people would.

There may also be some better biological options for supplying crops with phosphorus. *Mycorrhizal fungi* take up soil phosphorus (including some chemical forms not readily available to crops) and deliver it to roots. Macadamia nut trees and a few other crops produce short-lived clusters of *proteoid roots* (named for the plant family in which they were first described) that pump chemicals into the soil to solubilize phosphorus.[192]

In contrast to nitrogen-fixing rhizobia, however, mycorrhizal fungi and proteoid roots do not increase the total amount of phosphorus in the soil. Therefore, they are at most a partial solution to meeting crop phosphorus needs.[37] They may speed the recycling of phosphorus from the remains of previous crops and they may increase the thoroughness with which crops can use external phosphorus inputs, but they are not a long-term substitute for those inputs. This conclusion follows directly from the law of conservation of matter. So we might need to pay the energy cost of transporting manure back to farms to meet their phosphorus needs, even if there are other solutions to the nitrogen problem.

How do natural ecosystems deal with the phosphorus problem? Because natural selection improves genes and individuals rather than ecosystems, we should not necessarily expect natural ecosystems to distribute or "cycle" nutrients optimally. We should, however, expect sophisticated individual adaptations for nutrient acquisition. Proteoid roots are a good example. Why do proteoid root clusters have such short lifetimes? One possible explanation is that roots from neighboring competitors would grow into any proteoid root cluster that was active for a longer period, taking advantage of the solubilized phosphorus without paying the cost of making phosphorus-solubilizing chemicals. In some soils, proteoid roots mainly retrieve phosphorus from recently fallen leaves. Some plant would take up that phosphorus fairly soon anyway, but a plant with proteoid roots may grab it before another plant can.

Plants that acquire nitrogen or phosphorus through *symbiosis* (a close, long-lasting interaction between individuals of different species) with rhizobia or mycorrhizal fungi, respectively, have a number of sophisticated adaptations that increase their fitness on low-fertility soils. Rhizobia and mycorrhizal fungi also have numerous fitness-enhancing adaptations. But does natural selection among rhizobia favor traits that benefit their host plants? Not necessarily, as will be explained in chapter 9, where I discuss how an evolutionary perspective can help us improve symbiotic interactions with crops. Similar issues arise with mycorrhizal symbiosis.

## Nutrients, Refuging, and Ecological Footprints

Local sourcing of nutrients raises some additional issues, which I will illustrate with a brief digression. Heron Island is located near the southern tip of the Great Barrier Reef. When I got an invitation to address the Applied Evolution Summit there, the location was a significant plus. I wasn't disappointed with the research presentations,[193,194] or with the location. I got up in the middle of the night to watch green and loggerhead sea turtles laying eggs on the beach. During breaks, we went snorkeling on the reef, seeing lots of interesting corals and fish, a giant clam, and one large shark, apparently waiting in ambush under the shade of protruding corals. My dive buddy, Carl Bergstrom,[195-197] had swum within three feet of the shark's lair, a few minutes before the shark moved in.

The science fiction writer Arthur C. Clarke spent some time at the Heron Island research station and set part of his 1957 novel, *The Deep Range*, there. It's an interesting read. I'm not sure a book whose hero herds whales for meat would sell well today, even though he does eventually lead the conversion to dairy whaling. Today, the research station, where I stayed, shares the tiny island with a small national park and an expensive resort, and with thousands of nesting seabirds. Every tree branch seemed to have a noddy tern nest or two, each occupied by a hungry and demanding chick. Buff-banded rails, supposedly very shy in most places, paraded their chicks along the paths. At night, when the mutton birds—don't ask—started wailing, I had to wear earplugs to sleep. How could such a small island possibly produce enough food for so many birds? It couldn't, which brings me to the point of this little excursion.

Many people are actively committed to reducing their own negative impacts on the environment. *Ecological footprint analysis*[198, 199] has been a popular tool for analyzing those impacts. The idea is that if everyone consumed as many resources as the average American, we would need a bigger planet. Individuals' ecological footprints can be estimated as the land needed to grow their food, plus the land needed to grow trees to make the paper they consume, plus . . . well, then it starts to get complicated.

If we use electricity, how much land do we need to offset the negative effects of the sulfur dioxide emitted by the power plant? I could probably come up with a number, but would another scientist's estimate be anywhere close to mine? Not necessarily. Natural processes will eventually remove the sulfur dioxide from the atmosphere, but how many acres of golf courses or botanic gardens do we need to compensate us for the health effects it causes first? Sometimes, land area may not be the most useful measure of environmental impact.

And what about human reproduction, which is usually omitted from these calculations? One study added up the impact of subsequent generations, dividing environmental impacts among an increasing number of ancestors with each passing generation, and concluded that "each child adds about 9441 metric tons of carbon dioxide to the carbon legacy of an average female, which is 5.7 times her [direct] lifetime emissions."[200] Presumably, there were some males involved as well. I don't think most people would like to see humans die out altogether, however, or even fall to such low numbers that we can no longer maintain universities, railroads, the Internet, or even useful skills like sewing and fishing, which were apparently forgotten by the small and isolated aboriginal population of Tasmania.[201]

Despite these complications, there are some situations where ecological footprint analysis based on land area really does makes sense, which brings us back to Heron Island. To calculate the resource needs of birds on a per-acre basis, what land area should we use? Actually, we wouldn't use *land* area at all. The relevant areas are those in the ocean where the birds catch the fish they eat and feed to their chicks. This is a much larger area than the island itself.

Such situations, where resources from a large area are used in a small area, have been termed *refuging*.[202] And refuging, I suggest, is particularly relevant to agriculture.

Consider irrigation. Suppose a farmer applies twice as much irrigation water each year as falls on the farm as rain. That extra water often comes from somewhere else. In the central valley of California, much of the water used for irrigation originally fell in the Sierra Nevada Mountains as snow. Melting snow ran down to the valley in rivers or soaked into the ground and moved slowly through underground aquifers, before being pumped up through wells and used to water crops. So the ecological footprint of an irrigated farm should logically include the land where the snow fell. That same acre of mountain forest is doing other useful things at the same time, however, like absorbing carbon dioxide. So we probably shouldn't include the full area in the farm's ecological footprint.

And what about nutrients? How much should we add to a farm's ecological footprint to account for nitrogen fertilizer? Considering fertilizer as a potential pollutant, we should probably include some wetlands to denitrify (see glossary) excess nitrogen and some buffer strips to absorb excess phosphorus in runoff.[203] But what about the energy cost of making nitrogen fertilizer? If the energy comes from windmills, the land under the windmills can still be used for other purposes, including farming. What if the energy came from coal-burning power plants?

Calculating the ecological footprint for nutrients imported by organic farms may be simpler. If the nitrogen and phosphorus in the chicken manure applied to a 10-acre organic farm originally came from fertilizer applied to 40 acres of corn via the grain eaten by the chickens, does the 10-acre organic farm have a 50-acre ecological footprint, even without counting irrigation?

Maybe so, but there are some important points to keep in mind. First, it's not as if the manure was the only good produced by the chickens; their eggs fed a lot of people. Also, if the organic farm didn't take the chicken manure, that manure might have ended up polluting rivers or groundwater.

I once visited a family-run dairy farm in California with hundreds of cows. The farmer gave very detailed answers to all of our questions, until I (always a troublemaker) asked her about manure disposal. OK, they put the manure on their corn fields, but how many tons of manure did they apply per acre? I couldn't get an answer. If they were applying more nutrients in the manure than the corn could absorb—very likely, if the manure included nutrients from purchased grain and hay, as well as what they grew—then some of the surplus nutrients would percolate down through the soil. This would eventually pollute the aquifers that they, and their neighbors, depend on for well water. By taking some of that excess manure, organic farmers nearby could spread the nutrient resource over more land area, increasing the fraction that ends up in crops, rather than in the groundwater.

So my point isn't to criticize organic farms for turning a potential pollutant into a useful nutrient. This is a beneficial activity that should, perhaps, be rewarded with tax rebates or subsidies. But, if it takes 50 acres to grow enough grain to generate enough manure for 10 acres, that does raise the possibility that if *all* farms relied on manure for nutrients, we would need a bigger planet.

Or, perhaps, fewer people. My postdoctoral mentor, Bob Loomis, once analyzed the nitrogen budget of a fourteenth-century English farm.[11] More than half of the protein nitrogen in the wheat grain sold came from biological nitrogen fixation by legumes, but much of that occurred in meadows grazed by livestock. The manure they produced while penned up at night was collected and applied to the grain crop fields. The meadow-to-grain-field ratio was about 4:1, comparable to my assumption for organic farms. Even with this refuging of nitrogen from meadow to crop, wheat was typically grown only in alternate years. Additional nitrogen came from peas or leguminous weeds during the fallow year, yet yields were only about 1000 kg/ha, about a sixth of what can now be obtained every year with fertilizer. This system may have been sustainable, but producing grain on only 20 percent of the farm land, only in alternate years, and with a fraction of current yields could meet the protein and energy needs of only a much smaller population than we have today.

If Heron Island were the size of Tasmania, it could support huge populations of birds, but they would need more ocean to catch all those fish. Similarly, if all of agriculture were dependent only on refuging nutrients from larger areas into smaller areas, then supporting current human populations would require a much bigger planet. The fact that there isn't enough manure for everyone to farm organically, however, doesn't decrease the pollution-reducing contribution of current levels of organic farming. A mix of conventional and organic farms is also consistent with the bet-hedging principle.

To summarize the last two sections, we could greatly increase local sourcing of nitrogen through increased use of legume crops and forages that get their nitrogen from symbiotic, nitrogen-fixing rhizobia, but this would require major changes in farming practices and probably in our diets. For phosphorus, returning manure to where the animal feed was grown may be the only long-term option after phosphorus ores become scarce. This recycling will be difficult, however, so long as livestock-production areas (and, to a lesser extent, cities) are far from where feed is produced.

Local sourcing of nutrients in natural ecosystems—can we call the ocean around Heron Island "local"?—is a constraint imposed by the lack of external inputs, not an example of "nature's wisdom." But we may learn much from studying the adaptations of wild plants that evolved under that constraint.

## Are Mixtures the Best Way to Deploy Crop Diversity?

Jackson's and Piper's next suggestion was for greater reliance on "polycultures designed to benefit from spatial, seasonal, and nutritional complementarity among species."[14] A *polyculture*, or *intercrop*, is a mixture of two or more crops.

Two or more species may exhibit complementarity if they have different capabilities. For example, legumes can use atmospheric nitrogen (via rhizobial symbiosis), whereas grasses rely on soil nitrogen. Similarly, a cool-temperature and a warm-temperature species might produce more forage together, over the growing season, than either would alone. Mixed-species pastures are already common, for grazing livestock, but should we expand polyculture of crops for direct human consumption?

Although polyculture may sometimes be inspired by the plant mixtures often seen in nature, natural ecosystems dominated by single species are not rare. Redwood forests and estuaries dominated by eelgrass are two examples. Wild relatives of wheat, rice, and sorghum (all annuals!) grow in natural ecosystems with low plant diversity.[147]

Because natural selection hasn't optimized ecosystems, especially by agricultural criteria, it is entirely possible that these low-diversity natural ecosystems would actually be more productive, efficient, or stable if they were more diverse. By the same logic, however, the greater diversity found in other natural ecosystems does not prove that diversity enhances their ecosystem-level properties or processes, particularly those important to agriculture. Diversity may be there for a reason, just as earthquakes happen for a reason, but that does not mean that diversity is there for a purpose.

Fortunately, we do not need to rely on the false assumption that natural ecosystems have somehow adjusted their number of species up or down for optimal overall ecosystem function. Instead, we can refer to actual data on the effects of plant diversity in both natural ecosystems and farm fields.

The most common benefits attributed to species diversity are productivity and stability, so we will consider which kinds of complementarity increase productivity most. Intercropping may also increase stability, perhaps by lowering pest populations. That possibility will be discussed later in this chapter, so complementarity and productivity will be discussed first.

Experiments with wild species often seem to show greater productivity from mixtures, relative to the average of those species grown separately. One such experiment has been directed for many years by my ecologist colleague David Tilman, here in Minnesota. The higher-diversity treatments have indeed been more productive. However, as I will explain, it's not clear how well these results apply to other ecosystems, natural or agricultural.

Plots were seeded in 1994 with different numbers of wild prairie species per plot. By 1997, the eight-species plots had more than twice the growth of the average one-species plot.[204] But averaging across all one-species plots may not be the best measure of monoculture's potential. An experienced farmer would choose the best monoculture crops, not average ones. How did the high-diversity prairie plots compare with the best one-species plots? In 1997, only 10 percent of two-species plots had more growth than the best one-species plots, and plots with even more species did no better. Over several years, however, the performance of the more-diverse plots apparently improved. When results for 1999 and 2000 were averaged, almost half of the sixteen-species plots had more growth than even the best single-species plot.

## Possible Errors in Diversity–Productivity Experiments

A few years ago, my wife and I visited Cedar Creek, the site of this famous experiment. The more-diverse plots certainly had more growth per plot, but what struck me was the surprisingly large amount of bare soil in the one-species plots. The only data I found published on this were from 1996, when one-species plots had only 1/3 plant cover.[205] Low plant cover provides a simple explanation for why the one-species plots had such poor productivity: much of the sunlight potentially available to drive photosynthesis was hitting soil rather than leaves. Few farm fields—even few natural areas, aside from deserts—have so much bare soil by the middle of the growing season. Before applying results from Cedar Creek to agriculture, or to other natural areas, we need to know why the monoculture plots had such poor cover. Why didn't resident plants in the monoculture plots spread by runners or seeds into the bare areas?

Plots with a lot of bare soil might, perhaps, have had plenty of roots underground, consuming resources that seedlings would need to get established. But actually, less-diverse plots had less root mass and more soil nitrate than more diverse plots, so seedlings should have done at least as well as in more-diverse plots. Seedlings did germinate in the one-species plots. In fact, seedling biomass there was about ten times as great as in sixteen-species plots,[206] but they apparently didn't grow into adults. Why not?

One possible explanation is that plots were weeded three or four times a year, to remove seedlings not belonging in a plot. Weeders attempted to minimize disturbance by pulling weeds while still small, but could outnumbered resident-species seedlings have been accidentally damaged or removed along with the invaders? I also wonder whether human weeders might miss nonresident species more easily in a diverse plot than in a single-species plot. In a single-species plot, it may be more obvious which seedlings don't belong there, so they are more likely to be removed. It's interesting that a somewhat similar experiment nearby found "idiosyn-cratic" (that is, inconsistent) effects of species diversity on seed produc-tion, with diversity increasing seed production by some prairie species but decreasing seed production by others.[207]

Whatever the reason for the low plant cover in single-species plots at Cedar Creek, we know that agricultural monocultures usually achieve complete cover. Therefore, the poor performance of the one-species plots at Cedar Creek is probably not representative of most agricultural mono-cultures.

One place where we do sometimes see unusually low plant cover in agriculture is in the monoculture control plots of experiments designed to show the superiority of intercropping. For example, a photo in a 1982 *California Agriculture* article advocating polyculture shows a lot of bare space between plants in the monoculture plots,[208] reminiscent of the one-species plots at Cedar Creek. The polyculture plots, in contrast, appear to combine the total plant numbers from each of the monoculture treat-ments, leaving little bare soil.

To see the fallacy of this additive approach, take it to its logical ex-treme: an acre with one bean plant plus one corn plant would have higher total yield per acre than monoculture controls with one bean or corn plant per acre. But a more farmer-relevant comparison would be an acre of beans at their optimum density plus a separate acre of corn at its optimum density, compared to two acres of bean-corn intercrop at its optimum density.

Some intercropping researchers simply assume that some "recom-mended" density is optimal. However, this assumption needs to be tested at the actual research site. This is because the optimum density for the ex-cellent soils typical of agricultural research stations may be greater than recommended densities that are based on the average soils in a region.[172]

The Cedar Creek experiment is not the only study to show increased productivity from mixtures of wild species. In California, ecologists David Hooper and Jeffrey Dukes found that diverse grassland plots had higher productivity than the average single-species plot, except in the first of three years reported.[209] But perennial bunchgrasses alone outyielded more diverse plots (with bunchgrasses, legumes, and early- and late-season an-nuals) in two of three years. This seems to be the general pattern: diverse plots are usually more productive than the average single-species plot, but not necessarily more productive than the best single-species plot.

A farmer using monoculture would try to grow the best crop species, not an "average" species. Similarly, a farmer who chose to intercrop would look for the best combination. What combinations are likely to do best? This takes us back to the concept of complementarity.

## What Combinations of Species Show Complementarity?

*Spatial* complementarity would exist if a mixture of short and tall species, or a mixture of shallow- and deep-rooted species, used resources more efficiently. Above ground, the main resource is sunlight. Even short crops can use essentially all the available sunlight, so complementarity based on height seems unlikely, except perhaps in intercropping experiments where monoculture controls are grown at suboptimal densities.

Below ground, there must sometimes be water accessible to deep but not shallow roots. But deep-rooted plants don't necessarily ignore soil water near the surface. I have been unable to find any well-documented and well-managed experiments showing that a mixture of species grows better than the best species alone, solely due to differences in rooting depth. One experiment in Kenya found no difference in water use between annual crop monocultures in rotation (cowpea then corn) versus a polyculture mixture of those crops with a perennial hedge (*Senna spectabilis*).[210] The polyculture had less than half the grain yield of the monocultures. However, hedge clippings provided an additional benefit, as animal feed.

*Temporal* complementarity is a bit more promising. If the growing season is long enough, then farmers may grow one crop in the spring, followed by another in autumn. This only qualifies as polyculture if the two crops overlap in time. After corn plants have transferred leaf nitrogen to grain and dropped many of their leaves, but before the grain has dried enough to harvest, there may be enough light reaching the ground for photosynthesis by a second crop, like bean. A variation on this approach is sometimes used where a crop like cassava takes a year or so to reach full size, leaving some room for a faster-maturing crop in the meantime. I once saw an interesting version of this in a walnut orchard, where squash was grown for a few years between the young trees, until they got too big.

Growing an early- and late-maturing cereal together, in alternate rows, can sometimes give more total grain yield than growing them separately, but only if differences in timing are large. Research in Nigeria found no yield benefit of intercropping if the harvest dates for the two crops differed by less than 40 days, but up to a 16 percent gain for crops harvested 80 days apart.[211] Would it be worth modifying mechanized harvest equipment, as might be needed to harvest intercrops, for this relatively small yield increase?

These examples of temporal complementarity are based on two-species mixtures. If one wanted to mimic the seasonal complementarity seen in prairies, then one might use more than two species, probably including some perennials with spring growth and others with autumn growth. But harvesting seeds from different species over a period of months, without harming the other species, could be difficult and expensive. Again, the best harvesting device might be a goat or bison.

What about *nutritional* complementarity? Here, the best examples are mixtures of nitrogen-hungry grasses with nitrogen-fixing legumes.[212, 213] Where there is little available nitrogen per cubic centimeter of soil, a crop like corn needs many cubic centimeters of soil to get enough nitrogen. This requires such wide spacing between plants that a lot of sunlight hits soil rather than leaves. In that case, a legume crop that doesn't compete much with the corn for nitrogen may be able to photosynthesize using sunlight that would otherwise be wasted, assuming that there is enough water and phosphorus for both species.

Research in Kenya compared corn alone (with or without moderate amounts of nitrogen fertilizer) to corn intercropped with hedgerows of a legume shrub, *Leucaena leucocephala*. Nitrogen fertilizer increased corn yield often enough to be profitable for farmers. Therefore, we might expect a similar benefit from nitrogen-rich legume prunings, whether applied to the soil or fed to oxen that then supply manure. But like *Senna spectabilis,* discussed earlier, these legume hedgerows actually reduced the yield of corn, apparently by competing with it for water. However, the economic value of oxen feed reportedly outweighed the yield loss of the corn.[214]

In Denmark, a pea-barley intercrop yielded more than either barley alone or the average of peas and barley.[215] Peas alone was the highest-yielding alternative, however. This seems to be the usual pattern: most intercrops yield more than the average of the two crops, but less than the best crop alone.[216]

## Can Intercropping Increase Stability over Years?

Average yield is not our only criterion, however. Stability over years is also important. The idea that more-diverse ecosystems are more stable has a long history. Ecologist Daniel Goodman reviewed the information available in 1975 and concluded that "expectations of the diversity-stability hypothesis are borne out neither by experiments, by observation, nor by models . . . and its preconceptions are inconsistent with an evolutionary perspective."[217] However, he predicted that "the 'hypothesis' will persist for a while as an element of folk-science" among environmentalists "where it will doubtless do much good."

Today, the scientific status of diversity-stability is somewhat complicated. A recent review by ecologists Anthony Ives and Stephen Carpenter points out that species diversity may have contrasting effects on different forms of stability.[218] An ecosystem with more species may be *less* stable, in terms of susceptibility to secondary extinctions. That is, if one species goes extinct, the number of other species going extinct as a result tends to increase with the number of species and their interactions. On the other hand, more-diverse systems have greater opportunity for compensation, in which a surviving species provides similar functions as a species that went extinct. For example, if an ecosystem depends on nitrogen fixed by legumes, then having more than one legume species decreases the chance that a single species extinction will lead to nitrogen deficiency.

In agriculture, this sort of compensation-based stability could certainly be important for pastures of perennial species, where some of the nitrogen fixed by a legume this year may be available to grasses in future years. Legumes don't usually release much of their nitrogen into the soil until they die, however, so the nitrogen status of annual crops like wheat might not be improved immediately by growth alongside nitrogen-fixing vetch and/or peas. We had a dramatic illustration of this at LTRAS. *Volunteer* wheat plants (from seed dropped by the previous year's wheat crop) that came up in our nitrogen-fixing green manure crop often turned yellow, a symptom of nitrogen deficiency. This wheat looked even worse than in our unfertilized-wheat control plots. So the ability to fix nitrogen doesn't mean that a legume will share its nitrogen within a growing season, or even that it will compete less for soil nitrogen. Stability based on complementarity occurs in nature and sometimes in agriculture, but it's not universal.

## How Should We Use Crop Diversity to Control Pests?

Another of Jackson's and Piper's specific suggestions is to use "diverse cropping systems that encourage biological management of herbivores, weeds, and diseases."[14] Is this another possible benefit of intercropping?

I heartily endorse greater crop diversity, for various reasons. But should we always deploy that diversity in ways that resemble natural ecosystems? Many natural ecosystems (especially those dominated by perennials) have similar mixtures of species every year. Is that optimal for pest control? Or might diversity over time, like crop rotation, be better?

Can we at least assume that a mixture of two crops will have fewer pest problems than a single crop? In research done at Cornell University while I was a graduate student there, entomologists Stephen Risch, David Andow (now a colleague here at Minnesota), and Miguel Altieri reviewed 150 studies of the effects of plant diversity (intercropping or poor weed control) on the abundance of various insect pests. In about half the cases, insect pests were significantly less abundant in the more-diverse system, whereas more-diverse systems had significantly more pests only 18 percent of the time.[171]

So far, so good, although I worry that people who set out to study benefits of diversity may be less likely to publish if they find no such benefit. I'm not accusing anyone of a deliberate "cover-up." It's just that people who have some doubts about their results might be less motivated to do the additional work needed to publish them. This is why it's been suggested that drug companies should have to report results of all their trials, not just the ones that come out positive.

Could there be negative effects of diversity as well as benefits? For example, what if high crop diversity confuses pollinators as well as pests? We really need to know how effects of diversity on pests translate into effects on yield and its variability over years. Surprisingly, only 19 studies of 150 measured yield. Yields increased with plant-species diversity in 4 cases, but decreased in 9.[171]

Remember that lower yield means having to use more land to grow the same amount of food, perhaps leading to clearing forests or draining wetlands. Also, because irrigation requirements per acre don't change much with yield, a high-yield crop will use less water, relative to the amount of grain produced. In other words, higher yield often translates into higher water-use efficiency (WUE) as well. A high-diversity field with fewer pests but lower yield reminds me of a surgeon claiming "the operation was successful, but the patient died."

The review paper just discussed also found that only one study of 150 checked to see how much damage the pests—those that were often less common in more-diverse plots—actually caused. In this one study, it turned out that the yield loss was only 5 percent.[171] Given the technical challenges of managing and harvesting a mixture of two or more crops (even if they are grown in alternate rows, for example), I would want stronger evidence than this before recommending polyculture as a universal solution to insect pest problems. It may work well for some crop combinations, however.

What about disease? Fungal pathogens may spread more slowly through a field if only a fraction of plants are susceptible. For example, researchers working in China reported that intercropping helps to control a particular fungal disease. Instead of a random mixture of species, like those in natural ecosystems, they used rows of two different rice varieties in a specific pattern: four rows of disease-resistant rice between every two rows of disease-susceptible rice. The disease-susceptible rice was included because farmers get a higher price for that "sticky rice" variety, used in desserts. With greater distance between rows of sticky rice, losses to disease were reduced. With this intercropping approach, they could grow 18 percent as much of the sticky rice as if they had grown it in monoculture, plus about as much regular rice as their monoculture control.[219]

Getting as much regular rice as with monoculture, plus some sticky rice, seems like a good deal, if harvesting the intercrop isn't too tricky. But wait! Was the monoculture control grown at its own optimum density? As in the 1982 *California Agriculture* article discussed earlier, apparently not. A diagram in the article shows 30-centimeter gaps in the regular-rice monoculture control. In the intercrop, half of these gaps got sticky rice. Wouldn't filling in those gaps with regular rice instead, increasing the number of plants by 33 percent, have increased yield? If so, then their monoculture control isn't representative of real-world monoculture.

Now let's return to the question that opened this section: when we use crop diversity to control pests and disease, are "patterns and processes discernible in natural ecosystems"[14] really the most appropriate standard? Crop rotation, with different crops grown in successive years, is another way to use crop diversity, but it's very different from the patterns typically seen in natural ecosystems. Yet rotation has a long record of success.[220–222]

## Deploying Crop Diversity in Time versus Space

Crop rotation is, to some extent, an alternative to intercropping. Here's why. Suppose you want to grow corn, beans, and squash. There may be advantages to mixing all three crops together, as a polyculture. These benefits could include nutritional complementarity, discussed earlier, or reduced disease. In monoculture, a fungal spore released from a diseased corn plant early in the season is likely to land on another corn plant, where the fungus can reproduce and spread farther. In a mixture, many of the spores produced on a corn plant will land on a bean or squash plant, where corn-specific fungi can't reproduce. So diseases often spread more slowly through crop mixtures.

But with intercropping, what do you do in the second year? You've already used all three crops. Do you grow all three crops together again, year after year? Where corn is grown every year, any corn-specific pest or pathogen that can survive in the soil will tend to build up over years. With a three-year, three-crop rotation, on the other hand, a given field would have corn only once every three years, reducing buildup of corn-specific problems. With six different crops, you could have a three-year rotation of two-crop mixtures, of course, but you could also have a six-year rotation with one crop at a time. Which would have fewer pest and disease problems, over the long run?

I have never seen a good, published experiment to answer this question. One-year comparisons of intercropping and monoculture are common, but they don't tell us how intercropping compares with crop rotation over longer periods. The key comparison would be to grow the same intercrop repeatedly, while going through a rotation of the same crops at least twice. For three crops, that would mean a time commitment of at least six years. If the experiments were done by a graduate student, six years (plus a year or two of planning and analysis) is a long time to spend in graduate school. That may explain why such comparisons have apparently not been done.

While we wait for those experiments, computer simulations may be informative. Xiangming Xu, affiliated with research institutes in China and the UK, simulated different ways of deploying crop genetic diversity in space and time.[223] His model assumed that two varieties (each susceptible to a different fungal pathogen) were grown in a field divided into square blocks. The length and width of each block was assumed to be 40 times the average spore-dispersal distance. Disease development was compared among three cropping systems, which I will call *monoculture* (all blocks contain the same variety every year), *intercropping* (each block contains the same mixture of two varieties every year), and *crop rotation* (two varieties arranged like the two colors on a chessboard, with the varieties switching positions in alternate years).

With only one variety per block (monoculture or rotation), disease would spread more rapidly *within* blocks, relative to intercropping. But, with the chessboard arrangement (rotation), adjacent blocks would contain different varieties, creating major barriers to spreading *between* blocks. Furthermore, with crop rotation, many of the spores that over-winter would emerge the next year in a block with a nonsusceptible host.

Figure 7.1 shows the predicted spread of disease in successive years. Continuous monoculture, planting the same variety over the whole field each year, led to the highest disease levels, as we would expect. What about intercropping versus rotation? In the first growing season, intercropping gave the lowest disease levels. But the benefits of rotation became apparent in year 2, with lower disease levels than in the intercropping treatment. Results varied in subsequent years, because certain variables were assumed to vary randomly, as in the real world.

I wouldn't necessarily conclude that crop rotation always controls disease better than intercropping does. Simulation results depend on variables like the size of the blocks and the average distance traveled by fungal spores. But this analysis does confirm my suspicion that one-year comparisons of intercropping and monoculture can underestimate the merits of crop rotation. Models like this could help us decide how best to deploy crop diversity in space and time, keeping in mind the practical challenges of harvesting mixtures of species or varieties that may mature at different times.

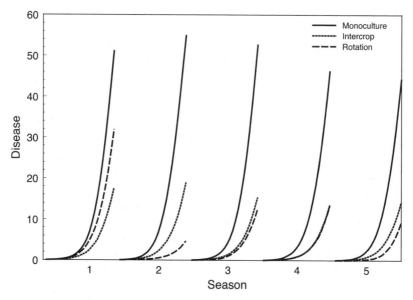

Figure 7.1. Predicted disease spread in successive years, with monoculture, intercropping, or crop rotation, based on model predictions from Xiangming Xu.

So let's increase crop diversity, but how should we use it? If we base our deployment of crop diversity on nature's wisdom, should we focus on intercropping or crop rotation? Natural ecosystems often have mixtures of plant species, as in intercropping, whereas they rarely have wholesale species replacement in alternate years (such as bundleflower in even years and big bluestem in odd years), like we see in crop rotation. I hope, however, that the previous chapter convinced most readers that patterns at the ecosystem level are not necessarily a good model for agriculture, because they have not been tested competitively by natural selection. So we should choose between intercropping and crop rotation (or some combination) based on the best available data and our own analysis, rather than slavishly copying the patterns seen in nature.

But wait—natural selection wants to comment on this question. Natural selection hasn't improved the overall organization of natural ecosystems, but it *has* tested and improved the defenses of individual plants. Do those defenses resemble intercropping, crop rotation, or some combination?

Some individual-plant defenses have an interesting resemblance to crop rotation. For example, many trees have *mast seeding*, with low versus high seed production in alternate years, often synchronized among trees. This can be an effective defense against seed-eating pests.[224] Relatively few seeds are produced in "off" years, which may somewhat limit populations of seed-eating pests. In "on" years, however, trees collectively produce more seeds than the seed eaters can consume. Synchronized *alternate bearing* in cultivated pistachio trees may provide similar benefits to mast seeding in wild tree species.[95]

Periodical insects—13- and 17-year cicadas are the best-known examples—appear to benefit similarly from emerging in such large numbers that predators can eat only a small fraction of them. There are also many periodical insects with much shorter, 2-year cycles. Often there are odd- and even-year populations, which would seem to undermine this pest-defense strategy. However, these two populations usually have little geographic overlap.[225] In any one location, therefore, emerging adults are abundant only in alternate years.

But a predator with a 2-year life cycle, synchronized with its prey, could perhaps be assured of abundant food. If the prey life cycle were 6 years, then predators with either 2- or 3-year life cycles could be a significant risk. No predator, however, is likely to specialize in prey with a 13- or 17-year life cycle. As in public-key cryptography,[226] large prime numbers are the key to security. Should agronomists be developing 13-year crop rotations?

Are mast seeding and periodic life cycles of insects really the sort of individual adaptations I have claimed we can expect from natural selection? Predator satiation looks like a collective benefit, in which some individuals are sacrificed to predators for the common good. In this and other perplexing cases, I have found it useful to ask, "ignoring group-level benefits, would a mutant with a different phenotype be more likely to flourish or to die out?" In other words, is this an evolutionarily stable strategy (see glossary)?[168, 227]

Imagine an individual pistachio tree that divided resources for reproduction equally among years, rather than using them mainly in years when other trees are "on." Any seed predators with a liking for pistachio nuts would focus on that one tree, in years when the other trees were "off." So mast seeding may provide a collective benefit, but its evolutionary persistence depends on its individual benefit.

## Perspective

Some of the specific suggestions made by Jackson and Piper, based on "patterns and processes discernible in natural ecosystems"[14] may deserve additional research as part of a diverse research portfolio, as advocated in chapter 12, but I have significant reservations about each of them.

Perennial crops can have various benefits, including erosion control. Perennials may sometimes use available resources, like sunlight and water, more completely than annuals do. However, evolutionary trade-offs between reproduction and longevity (or seed growth and root storage, as constrained by the law of conservation of matter) suggest that it will be impossible to approach the high grain yields of annual crops without losing perenniality. It may therefore make more sense to use wild or domesticated perennials as forage for grazing animals, perhaps including bison.

Conservation of matter also limits our ability to meet crop nutrient demands entirely from local sources, although this may be possible for nitrogen, if we increase our use of legumes supporting nitrogen-fixing symbiotic bacteria. My pessimistic view of local nutrient recycling assumes that many people will continue to live in cities. Greater local reliance for nutrients could be possible if people (and also especially farm animals) were dispersed much more evenly across the landscape. The limited ability of natural selection to optimize ecosystem-level nutrient cycling is an additional constraint, although natural selection has given individual plants some sophisticated adaptations for nutrient acquisition, such as proteoid roots.

The extent to which the productivity of natural ecosystems depends on the number of species present, as opposed to which species are present, is unclear. Much of the research on mixtures of crop species or genotypes (polyculture or intercropping) has been plagued by errors, such as suboptimal densities in monoculture controls, or not continuing the experiment long enough to compare the intercrop with a rotation of the same species. Although there are well-documented yield benefits from mixing legumes and nonlegumes on nitrogen-poor soil and from pairs of crops with only limited overlap in time, such benefits rarely exceed 20 percent. This may be enough to justify intercropping in some situations, but it is a small increase relative to the doubling or more obtainable with nitrogen fertilizer and/or improved varieties.

Given the practical difficulty of managing mixtures, it is not surprising that organic farms I have visited rarely use intimate mixtures of species. Mixtures of salad greens are one common exception, motivated by convenience in harvest and marketing rather than by any particular complementarity during growth. The most widely used intercrop worldwide, by far, is a mixed grass/legume pasture grazed by animals. If perennial crops are needed to control erosion, then we may need to restore grazing animals to their ancient role in nature and in agriculture.

I strongly support the call for more diverse cropping systems, but not necessarily ones that resemble natural ecosystems. Haphazard ecosystem-level processes may have led to plant communities resembling polyculture, with little variation in species composition over years, but there is no reason to expect this pattern to be optimal. Individual-based natural selection has given us strategies that more closely resemble crop rotation, with diversity in time rather than in space. Mast seeding and the 13- and 17-year periodic cicadas are examples of strategies driven by individual selection but benefiting entire populations.

Sometimes, however, collective benefit and individual benefit are in conflict. Such conflicts have limited the evolution of cooperation within and among species. Although cooperation is very widespread in nature, it may need some help from humans to achieve its full potential. Enhancing cooperation is an important theme for the rest of this book, starting with the next chapter.

# 8

## What Has Worked

IMPROVING COOPERATION WITHIN SPECIES

UP TO NOW, I have emphasized the limitations of two popular approaches for improving agriculture. In chapter 5, I suggested that much of the current effort in biotechnology will fail, mainly because it ignores tradeoffs. In chapter 7, I was almost as critical of copying the overall organization of natural ecosystems. The rest of this book has a more positive tone, discussing approaches that have worked in the past, as well as those with unrealized potential.

### Beyond the Green Revolution

While rejecting the extreme positions taken by some advocates of biotechnology and some advocates of "farming in nature's image," I am not advocating splitting the difference. That approach rarely leads to optimal solutions.[228] Instead, let us take a multidisciplinary approach, with evolutionary biology at the center.

Despite my earlier criticisms, I recognize each group's expertise. Biotechnology may well speed our progress, once we identify realistic goals. And we can certainly learn much from studying the organization of natural ecosystems, so long as we carefully analyze what works and what doesn't, rather than simply assuming perfection. Detailed knowledge of the inner workings of plants and of the workings of natural ecosystems are both likely to be useful, but we can no longer afford to ignore evolution.

We need to pay particular attention to evolutionary tradeoffs, especially tradeoffs between individual-plant competitiveness and the collective performance of plant communities. Opportunities linked to these tradeoffs are the theme of this chapter.

Natural selection favors individual competitiveness over community performance, whenever the two are in conflict. So we can sometimes improve whole-crop efficiency by breeding for cooperation among plants, sacrificing individual-plant competitiveness. Implicit acceptance of this tradeoff has been key to many past improvements, but we can make faster progress by recognizing tradeoffs explicitly.

As I was writing the first draft of this chapter, newspapers reported the death of Norman Borlaug, the plant breeder who was awarded the Nobel Peace Prize in 1970 for his leading role in the Green Revolution. When a magazine cover referred to "the man who saved a billion lives, but you've never heard of him," I knew immediately who they meant. Of course, many people contributed to the Green Revolution, developing improved rice and wheat varieties and crop-management practices which, together, doubled or tripled yields.

Like many other plant breeders before and since, Borlaug was successful in developing varieties with better disease resistance. But Borlaug's most-lasting contribution has almost the opposite relationship to evolution as his work on disease resistance. Plant breeders increase disease resistance by accelerating evolution that would happen anyway, albeit more slowly, by natural selection. Borlaug's major contribution, in contrast, was based on reversing past natural selection, although I don't know if he saw it that way.

In 1953, he started working with short-stemmed (that is, intrinsically less competitive) wheat varieties developed by University of Washington plant breeder Orville Vogel. Vogel had started with wheat variety Norin 10, which was developed in Japan by Gonjiro Inazuka and collected there by Cecil Salmon, who was working with the U.S. occupation forces after World War II.

Popular discussions sometimes characterize the Green Revolution as "using more fertilizer," but that is a major oversimplification. Conservation of matter implies that a farm exporting more nitrogen in grain protein will indeed need more nitrogen inputs, from some source. But just adding more fertilizer to old crop varieties wouldn't have increased yields much, if at all. This is because older, taller varieties were unable to support heavier heads of grain without *lodging*, or falling over. Lodging may be directly caused by wind, but taller varieties and plants top-heavy with grain are much more susceptible. At LTRAS, for example, we rarely saw lodging in our unfertilized control plots, where nitrogen deficiency limited stem growth.

Shorter Green Revolution varieties made it possible to use more nitrogen fertilizer, increasing grain growth without lodging. Reduced lodging is not the main reason that shorter varieties have higher grain yield, however. A field of shorter plants can capture as much sunlight as a field of taller plants, so photosynthesis rates are similar, but they use more of their photosynthate to make grain, rather than stem.

Why did past natural selection make wheat plants (and their wild ancestors) waste resources on tall stems, at the expense of seed production? Surely an annual plant that produces more seeds will have more descendants than one producing fewer seeds, all else being equal. Figure 3.1 offers an explanation. The wild ancestors of wheat grew as mixtures of different genotypes. Shorter plants put a greater fraction of their resources into seed growth, but they were shaded by their taller neighbors. Taller plants therefore photosynthesized more, increasing their individual seed production relative to shorter plants, despite their greater investment in stem growth. Thus, past natural selection for individual competitiveness resulted in genotypes that were relatively inefficient in producing seeds.

Natural selection continues today, whenever plants of different genotypes compete, and it still favors taller genotypes, even if they are lower-yielding. A clear example comes from experiments at the International Rice Research Institute. A shorter rice variety developed during the Green Revolution had much higher grain yield when grown alone than an older, taller variety, mainly because the short variety invested more in grain and less in stem. But what happened when the two varieties were mixed 50:50?

Each year, a random sample of the seed produced by the mixture was replanted, mimicking competition in nature. In only three years, the high-yield variety had disappeared, outcompeted by the low-yield variety.[229] So there was a tradeoff between individual-plant competitiveness and collective productivity.

## Reversing Evolutionary Arms Races

The tendency of natural selection to favor taller plant genotypes, thereby diverting resources to stems at the expense of grain, is an example of a wasteful evolutionary arms race.[230] Two neighboring countries that fear each other may each invest so much in weapons that they drastically reduce their ability to meet other needs, while leaving the balance of military power unchanged. Similarly, a wheat or rice population that evolves increasing height, in response to competition with neighbors, will inevitably invest less in grain. Yet there is no photosynthetic advantage to increased height when neighbors are taller, too (see figure 3.1). What is the optimum height? A little taller than your neighbor.

One difference is that humans can negotiate arms-control agreements that increase security while reducing wasteful military spending. Plants never negotiate arms-control agreements, as far as we know, but humans can sometimes intervene on their behalf. We can identify cases where past evolutionary arms races have undermined the collective performance of plant communities, and then impose selection for more "cooperative" genotypes.

Selection for shorter, less-competitive, but higher-yielding plants during the Green Revolution is the best-known agricultural example of reversing an evolutionary arms race, but it is not the only one. Evolutionary trade-offs also occur underground. Natural selection favors rooting patterns that steal soil water and nutrients from under neighboring plants, collectively wasting resources to make extra roots that don't increase the total water and nutrient uptake by the plant community.[231] This is another example of an arms race, where plants evolved increasing investment in aggressive root competition, leaving them with fewer resources to invest in seeds than if they showed more mutual restraint. Again, plant breeders could reverse the effects of past natural selection by breeding for roots that are less prone to invade the soil beneath neighbors.

## Colin Donald: From Competitiveness to Cooperation

A comprehensive approach to reversing the negative effects of past evolutionary arms races on the collective performance of crops was proposed in a landmark paper by Australian agronomist Colin Donald.[112] The paper is most often cited for inventing the term *ideotype*, a plant design based on our understanding of traits that affect yield.

But Donald's key points involve the tradeoffs that make evolutionary arms races so counterproductive. He argued that genotypes with high yield potential will usually be less competitive against other plants. Achieving their greater yield potential may depend on optimal spacing or better weed control. Changing spacing between plants can even change the relative performance of different genotypes; those that yield more grain per plant at wide spacing often yield less at the closer spacing that maximizes total yield.

What are some traits that can increase whole-crop yield at the expense of individual-plant competitiveness? In addition to being short, Donald suggested, a high-yielding grain crop should have fewer and smaller leaves. At wide spacing, this reduced leaf area per plant would be insufficient to catch all the available sunlight. But when the plants are grown close enough together, even smaller leaves will completely shade the ground, capturing almost all of the available sunlight. And the resources not used to make extra leaves can be used to make more grain.

A plant with more leaves will tend to shade a neighbor with fewer leaves, so past natural selection has given us plants that make more leaves than are needed to capture all the available sunlight. But we can practice artificial selection for plants with less leaf area. To maintain this trait, we would need to limit natural selection for greater leaf area by not mixing genotypes differing in leaf area, at least in fields harvested for seed. We could still grow mixtures differing in other traits, such as disease resistance.

Donald also suggested that plants like wheat should have only one stem per plant. Most wheat genotypes produce multiple branches near the ground, which grow like stems. Unless plants are grown at very wide spacing, these stems spread out and invade the territory of neighboring plants. One problem with multiple stems is that a multistem plant doesn't start out with lots of stems. It starts as a little seedling with a few small leaves. It will eventually expand to fill the available space and catch all the available light, but "eventually" isn't good enough. Any photons missed early in the season are gone forever, and never contribute to photosynthesis or growth. With a large number of closely spaced single-stem plants, however, the available space will be filled sooner and fewer photons will be missed.

Donald also advocated more-vertical leaves. When the sun is more-or-less overhead, a vertical leaf will catch less sunlight than a horizontal leaf of the same size. This could decrease whole-crop photosynthesis, if plants are too widely spaced. If the plants are grown closely enough together, however, even vertical leaves will catch most of the available sunlight.

What is the advantage of vertical leaves? In most crops, photosynthesis shows diminishing returns with additional light. By spreading the available light over a larger leaf area, vertical leaves increase the illuminated leaf area more than they decrease photosynthesis per unit area. If all plants have vertical leaves, therefore, whole-crop photosynthesis increases.

In modern agriculture, there may be little genetic diversity within a field, limiting the potential for ongoing natural selection to increase height or make leaves more horizontal. So the wasteful competitive traits discussed here are mainly the result of past natural selection. These traits can represent opportunities for significant genetic improvement of crops—opportunities missed by natural selection.

To summarize, Donald hypothesized that crop genotypes with greater individual-plant competitiveness will, by wasting resources in competition and by suppressing the growth of fellow crop plants, have poor collective performance when grown together. He focused on yield, but similar arguments would apply to other important traits, like water-use efficiency (WUE).

## Testing Colin Donald's Hypotheses

Donald's classic paper already included evidence for the value of each proposed trait, mostly based on his own published research or that of others. Since then, a few people have conducted explicit experimental tests of Donald's hypotheses. For example, Japanese crop physiologist Makie Kokubun compared single-stem soybean plants (created surgically) to multistem plants.[232] At wide spacing, a community of single-stem plants had lower yield, presumably because some photons passed between the plants, rather than being captured for photosynthesis. With closer spacing, however, single-stem plants had higher yield, consistent with Donald's hypothesis.

It is also possible to test Donald's ideas using data collected for other reasons. One approach is to compare several genotypes and plot their yields as a function of the traits discussed by Donald. I have done this using data published by Roger Austin and colleagues, comparing twelve wheat genotypes developed between 1908 and 1978. A major conclusion was that increases in yield potential over that period were mainly due to genetic improvement in *harvest index* (that is, greater allocation to grain), rather than genetic improvement in productivity per se.[233] Plotting yields of the different genotypes against their harvest index values supported this conclusion. Much of this increased allocation to grain could be explained by the shorter height of the higher-yielding varieties.

There was little difference in overall plant growth between older and more-modern genotypes. This lack of improvement is consistent with the claim in chapters 4 and 5 that natural selection has left few tradeoff-free opportunities for humans to increase the efficiency of fundamental processes like photosynthesis.

The improvements proposed by Donald are not tradeoff-free, however, which explains why they were rejected by natural selection. Shorter plants with fewer stems and fewer leaves are less likely to shade their neighbors and more likely to be shaded by them. They may also require closer and more-uniform spacing to achieve their yield potential. For practical reasons, Austin and colleagues grew all genotypes at the same spacing. They may therefore have underestimated the value of some traits whose benefit depends on closer spacing. Nonetheless, their data show a clear trend for varieties with fewer stems per plant to have higher grain yield.

Incidentally, they also compared yields in two fields, with soils that were low versus high in nitrogen. In almost every case, the varieties that yielded most with high nitrogen also yielded most with low nitrogen. Similarly, Green Revolution rice varieties have higher yields than older varieties, whether or not they receive fertilizer.[234] A tradeoff between yields with and without fertilizer is sometimes assumed to exist, but this may not be true. Whether or not plants are nitrogen-limited, putting a higher fraction of total resources into grain can increase yields. Modern varieties often benefit more from fertilizer than older ones did, but it is not true that they "require" fertilizer to achieve any yield increase.

The modern varieties do probably require better weed control, however. The same traits that make them less likely to wastefully suppress growth of neighboring crop plants may also make them less competitive with weeds. The possibility of crops that are more competitive against weeds, without suppressing each other, will be deferred until later in this chapter.

Donald focused on wheat. What about corn? A sixfold increase in U.S. and Canadian corn yield between 1930 and 2000, partly due to nitrogen fertilizer and improved crop management, has apparently involved increased total growth, rather than increased harvest index.[235] This contrasts with the results of Austin and colleagues for wheat. As noted, however, their use of the same spacing for all varieties may have missed some increased growth potential resulting from interaction between genotype and spacing. The increased growth of modern corn varieties is partly due to more erect leaves,[84] as proposed by Donald and consistent with experimental manipulation of leaf angle in corn.[111]

Some of the increase in corn yield is apparently due to its photosynthesizing over a greater fraction of the growing season. And much of the increase in corn yield has been attributed to increased stress tolerance. Does this disprove my hypothesis that natural selection is unlikely to have missed simple, tradeoff-free improvements?

Not necessarily. The wild ancestors of corn lived in Mexico, and evolution of adaptation to colder climates has been occurring only since indigenous migration brought domesticated corn north. So there might be more room for improvement in corn's tolerance to low temperatures than to high temperatures, for example. Consistent with this hypothesis, photosynthesizing earlier is partly due to planting earlier, linked to greater cold tolerance. But photosynthesizing later is partly due to the "stay-green" trait.[235] Did older varieties transfer nitrogen from leaves to grain earlier, perhaps because they were grown in lower-nitrogen soils?

Modern corn varieties also seem to be more resistant to "plant population density stress."[235] That is, they tolerate crowding better. This has also been seen in wheat.[236] Both crops are grown closer together now than they were in the past, so it's not surprising that there was room for improvement in this trait.

But has drought tolerance of corn been improved by plant breeding? The wild ancestors of corn in Mexico were presumably exposed to drought, so I wouldn't have expected natural selection to miss tradeoff-free opportunities to improve this trait. Experts apparently disagree about whether plant breeding has increased the drought tolerance of corn.[237, 238] A more-recent variety may yield better under drought than an older one does, but is that because of an increase in drought tolerance, or some other improvement? In either case, was that improvement tradeoff-free? The best discussion I've seen of these issue comes from work on wheat, discussed in the next section.

## Drysdale Wheat, Tradeoffs, and Efficient Water Use by Crops

My tradeoff hypothesis might seem to imply that drought-resistant varieties would have lower yield than less-resistant varieties, under wetter conditions. Jeremy Burdon, an evolutionary biologist now in charge of plant research for Australia's main research agency, called my attention to a possible counterexample. An Australian wheat variety, Drysdale, has higher yield under drought than previous varieties, without suffering lower yield under wetter conditions.

But lower yield under wet conditions is not the only possible tradeoff. So I looked for published research on Drysdale wheat. What I found was quite impressive, but it didn't change my views about the importance of tradeoffs. Rather the opposite, in fact.

A paper by people involved in developing Drysdale wheat included yield comparisons under a range of conditions.[9] Drysdale outyields older wheat varieties by up to 40 percent at the driest sites. At the wettest sites tested, it does about as well as the older varieties. So, from a practical standpoint, this does look like a tradeoff-free improvement.

But the tradeoffs are there. Many of them are discussed, with admirable clarity, in the same paper that reports the lack of a wet-versus-dry tradeoff.[9] For starters, the authors recognized that mere *drought tolerance*— the ability to survive drought—wouldn't necessarily help much. Wheat needs to grow and produce grain, not just survive.

What we need is water-use efficiency (WUE), photosynthesizing more and producing more grain, while using less water. Water-use efficiency at the leaf level is the ratio of photosynthetic uptake of $CO_2$, divided by transpirational water loss from leaves.

Carbon dioxide diffuses into leaves through the same stomata through which water vapor diffuses out. Diffusion of each gas is driven by the difference in its concentration between the inside and the outside of the leaf. Everyday examples of analogous cases include conduction of heat through the wall of a house (driven by the temperature difference between the inside and outside) or flow of water through a pipe (driven by pressure differences between its two ends). Readers who have studied physics or electronics may remember Ohm's law, describing flow of electric current through a wire, driven by voltage differences. The math behind all of these is the same.

Combining the diffusion rates of carbon dioxide and water vapor, we get:

$$\text{WUE} = \text{photosynthesis} / \text{transpiration} = (Ca - Ci) / (Wi - Wa)$$

where $Ca$ and $Ci$ are $CO_2$ concentrations in the atmosphere and the leaf interior, while $Wa$ and $Wi$ are corresponding water-vapor concentrations. This equation simply says that water-use efficiency is the rate of $CO_2$ diffusion into the leaf (equal to photosynthesis rate; otherwise $CO_2$ would accumulate) divided by the rate of transpirational water loss. Carbon dioxide diffusion into the leaf is driven by the difference in $CO_2$ concentration between the atmosphere and the leaf interior. Similarly, water loss from the leaf depends on the difference in water-vapor concentration between the leaf interior and the atmosphere. (For simplicity, I have left out a constant that corrects for differences between $CO_2$ and water vapor, which doesn't affect any of my conclusions.)

So one way to increase WUE is to increase $Ca$, the $CO_2$ concentration in the atmosphere. Some greenhouse managers do this through $CO_2$ fertilization. For crops grown outside, $Ca$ is increasing over decades as we burn coal and oil. All else being equal, this increase in $Ca$ would increase WUE. But increasing $Ca$ can have various negative side-effects, including increased leaf temperatures—atmospheric $CO_2$ traps outgoing heat and reradiates it back to the earth's surface—which can raise $W_i$ and thereby lower WUE.

Another way to increase WUE is to decrease $Ci$. In theory, this could be done by increasing photosynthetic capacity, using up $CO_2$ inside the leaf faster. For example, higher-nitrogen leaves can have greater WUE. This is because a crop with plenty of nitrogen can make more protein, including rubisco, the key photosynthetic enzyme. So one way to increase photosynthesis, decrease $Ci$, and increase WUE, is to use nitrogen fertilizer. Most farmers in industrialized countries already use more than enough nitrogen fertilizer, however, leaving little room for additional improvement by this route.

What about genetic approaches? The Drysdale wheat paper notes that a crop with a given amount of nitrogen could decrease $Ci$ if it has alleles for making a few high-nitrogen leaves, rather than many low-nitrogen (low-N) leaves.[9] At the individual-leaf level, this would increase WUE. But a crop with fewer leaves won't capture as much sunlight for photosynthesis, at least at wide spacing. So there's our first potential tradeoff, lower whole-crop photosynthesis.

And that's not all. Sunlight not intercepted by smaller, high-nitrogen (high-N) leaves will instead hit the soil, increasing evaporation of water from the soil surface. The water-use efficiency of water evaporating from the soil is zero, because it makes no contribution to photosynthesis. So there can be a tradeoff between individual-leaf WUE (photosynthesis per transpiration), maximized by high-N leaves with low $Ci$, and farm-level WUE (grain production divided by total water use, including that lost to soil evaporation), increased with earlier shading of soil by thinner, low-N leaves.

Despite these possible tradeoffs, Drysdale breeder Richard Richards has made good progress developing plants with low $Ci$, and hence high WUE, using traditional backcross breeding. Fortunately, his high-school friend, Graham Farquhar, had discovered an efficient way to screen for low $Ci$.[239]

Most of the carbon (C) in $CO_2$ has an atomic weight of 12 (twelve times the weight of a hydrogen atom), but about 1 percent is the heavier C-13 isotope. Photosynthesis preferentially takes up C-12, so more than the expected 99 percent of C in plants is C-12, usually. But discrimination in favor of C-12 decreases at lower $Ci$; beggars can't be choosers.

So Richards could indirectly select for low $Ci$ (and hence high WUE) by choosing plants with relatively more C-13. The isotope measurements could be made using leaf samples from plants that later made seeds, which could be used for subsequent breeding. Drysdale wheat was the result of repeated back-crossing and selection based on the ratio of C-13 to C-12.

Why is the internal $CO_2$ concentration, $Ci$, lower in Drysdale wheat, and what are the implications for tradeoffs? That's not entirely clear. As discussed earlier, an increase in photosynthetic capacity would pull $Ci$ lower, thereby increasing WUE. But why would past natural selection have missed this "low-hanging fruit," an opportunity to increase both photosynthetic capacity and water-use efficiency?

One possible reason is that photosynthetic capacity depends on rubisco, so it depends on leaf nitrogen. Much of the evolutionary history of wheat took place in nitrogen-limited environments, so wheat might have inherited a conservative nitrogen-use strategy that may no longer be appropriate. Of course, what's appropriate today could depend on the economic and environmental costs of nitrogen fertilizer.

An alternative explanation for lower $Ci$ in Drysdale wheat is partial stomatal closure. A leaf with a given photosynthetic capacity will pull $Ci$ lower, if stomata limit $CO_2$ influx more. Limiting $CO_2$ influx implies lower photosynthesis rate, but transpiration can be reduced even more than photosynthesis, increasing leaf-level WUE. So there's another tradeoff, between photosynthesis rate and water-use efficiency.

You might expect a lower photosynthesis rate to translate into lower yield, yet Drysdale didn't have lower yield in any of the environments tested. The yield advantage of Drysdale, relative to a previously recommended variety, did decrease as rainfall increased, however, and it still seemed to be decreasing toward the wetter end of the range.[9] So maybe Drysdale would do relatively poorly under even wetter conditions than those tested. Even if that turns out to be true, however, it would not undermine the practical value of this work. This is because the range of conditions tested is representative of a wide variety of Australian wheat farms. Farmers don't care how a wheat variety would yield with more rain than they are likely to get!

Still, is it possible to have a lower photosynthesis rate, yet higher yield? Absolutely. Yield depends on total seasonal photosynthesis, not the rate of photosynthesis on any given day. If, by using water more efficiently, Drysdale can photosynthesize for more days before it runs out of water, then its total seasonal photosynthesis could be greater than a less-efficient variety with higher daily photosynthesis. Crop ecologists sometimes measure soil water as "days of transpiration."

But why would past natural selection have missed this opportunity to increase seed production, by using water more efficiently? This could be yet another example of the tradeoffs discussed by Colin Donald, between individual competitiveness and collective crop performance. An individual plant that sacrifices photosynthesis today, to save water in the soil for tomorrow, will be outcompeted by plants that maximize their own photosynthesis today, using the water the frugal plant saved. I predict that careful experiments would show that Drysdale is less competitive than the less-efficient varieties it replaced, but perhaps only slightly. That's a tradeoff, but probably one we can accept.

In some cases, the preceding equation implies even more dramatic opportunities to improve water-use efficiency, based on a tradeoff that constrained past natural selection but need not constrain us. This example is mentioned in the same excellent paper that discussed the development of Drysdale.[9] Water-use efficiency can be increased, even without increasing photosynthesis, by decreasing water loss to transpiration.

We have only limited control over leaf-interior water-vapor concentration, $Wi$, which depends mainly on leaf temperature. (For example, stomatal closure to reduce transpiration reduces transpirational cooling, increasing leaf temperature and therefore $Wi$, which may reverse the decrease in transpiration.) But what about $Wa$, also known as *atmospheric humidity*? Humidity tends to be higher early the morning, so there may be some potential to increase WUE through partial stomatal closure during the afternoon. Crops would then use proportionally more water in the morning, when the $CO_2$-to-water exchange rate is more favorable.

The big difference in humidity, however, is often that between summer and winter. Chickpea growers in Syria doubled their yields under water-limited conditions, just by growing their crop in winter rather than summer.[240] The humidity was higher in winter, so $Wi - Wa$ was less, increasing WUE and yield.

This may seem obvious, but it took some work. The problem was a fungal disease, prevalent in winter, that had previously destroyed the crop. Fortunately, plant breeders were able to develop a disease-resistant variety of this crop, as they have for so many others. Presumably, they will need to keep developing new disease-resistant varieties, as the fungus evolves its ability to attack the crop. Nonetheless, this example shows how creativity can find indirect solutions to problems that defy a direct approach. In this case, disease resistance made it possible to grow the crop in the winter, indirectly increasing water-use efficiency to a degree that no direct approach has yet matched.

## Solar Tracking: An Evolutionary Arms Race?

Are there more opportunities for crop genetic improvement, linked to other tradeoffs between individual-plant competitiveness and the collective performance of crops? Solar tracking by leaves may involve such tradeoffs. Many plants turn their leaves over the course of the day, keeping the flat surface facing toward the sun. Look at clover plants in a lawn, for example, and you will often see that they face east in the morning, south at noon—or north, in the southern hemisphere—and west in the evening. A leaf that tracks the sun catches more photons, so it will often have greater photosynthesis. But a solar-tracking leaf also casts a larger shadow, reducing the photosynthesis of leaves below or behind it.

To predict the overall effects of solar tracking on photosynthesis, I used a computer model of the alfalfa crop, which I developed years earlier under the guidance of my postdoctoral mentor, Bob Loomis.[241] With widely spaced plants (or with few leaves per plant), solar-tracking leaves mainly shade the soil surface rather than other leaves. At wide spacing, therefore, the model predicted an increase in photosynthesis from solar tracking.

*Leaf-area index* (LAI), the ratio of leaf area to ground area, is often used to predict the fraction of sunlight captured by a crop. (Because some leaf overlap is inevitable, even at low leaf area, LAI = 1 doesn't come close to 100 percent light interception.) One obvious prediction of our model was that, especially as LAI increases, solar tracking would increase the shading of lower leaves, relative to an alfalfa genotype that didn't track the sun.

But it matters which leaves get the sunlight. Because upper leaves already have plenty of light, they benefit less from each additional photon than a lower leaf would have. (The logic is similar to that for vertical versus horizontal leaves, discussed earlier.) Therefore, our model predicted that solar tracking would actually decrease overall photosynthesis of dense alfalfa canopies.

Some years ago, I set out to test this hypothesis with the help of two research technicians: Jim Fedders and then Barry Harter, at the USDA lab in West Virginia where I worked at the time. Every cloudless morning—these weren't as common as we would have liked—we measured the photosynthesis of a group of alfalfa plants under a clear plastic dome. In our first measurement of the morning, the plants were naturally tracking the sun. But then we turned the plants 180 degrees, disrupting solar tracking.

Analyzing our data proved to be more complicated than expected, but our overall findings were clear. When solar-tracking leaves mainly shaded the soil rather than other leaves—that is, at low LAI—disrupting natural solar tracking decreased photosynthesis. But the photosynthetic benefit of solar tracking was small, rarely more than 5 percent. Even this small benefit decreased as plants grew more leaves, so that leaves were more likely to shade each other than to shade the soil.[242] There were some cases, all at high LAI, where disrupting solar tracking actually increased photosynthesis.

However, the negative effects of solar tracking at high LAI were less than predicted by our computer model. Clearly, the model was wrong. But which model? The computer model, or our "model system," a 1-square-meter group of plants?

Our little group of plants (turned or not) was intended to be a model of a whole field of tracking or nontracking alfalfa. But, unlike lower leaves in the middle of a big field, the lower leaves of our model plants received light from the side, an example of an *edge effect*. This extra light available to lower leaves reduced the negative effect of shading by solar-tracking upper leaves. So it is possible that the computer model was right and the physical model was wrong. Additional experiments might be able to correct for edge effects and give a more accurate measure of the costs and benefits of solar tracking.

If solar tracking really does decrease photosynthesis overall, why hasn't it been eliminated by natural selection? This would be an example of a *simple* improvement (that is, one that has arisen frequently), because solar tracking is a complex process, dependent on hundreds of genes. Mutations that knock out some key element of this process must therefore arise fairly often. Why haven't nontracking mutants taken over?

To answer this question, we need to consider *whose* leaves are shaded by leaves that track the sun. Natural selection would tend to eliminate solar tracking, if individual plants usually shade their own lower leaves. But plants often grow so close together that a plant is more likely to shade its neighbor than itself, just as shadows from corn tassels may mostly shade neighbors. Solar tracking may reduce photosynthesis by a few percent, but it reduces light reaching the ground by 50 percent. Shading by solar-tracking leaves may therefore suppress seedlings that could otherwise grow into serious competitors for light and other resources. So solar tracking may be the result of yet another evolutionary arms race. If this is true, then we might be able to increase the productivity of alfalfa and other solar-tracking plants by eliminating this trait.

But remember the small photosynthetic benefit to solar tracking at low LAI, when solar-tracking leaves were more likely to shade the ground than each other. Seedlings that are too small and widely spaced to shade each other could benefit from tracking the sun, capturing more light for photosynthesis. There is also little mutual shading in alfalfa crops that are just starting to regrow after hay harvest. The most productive crop, therefore, might be one that retains the solar-tracking trait but turns it off as mutual shading increases. This might not qualify as simple, however.

## Can Cooperative Wheat Suppress Weeds?

The ideal crop would photosynthesize as efficiently as possible, while also shading and suppressing weeds. Is there a conflict between these goals?

Leaves use light more efficiently (more photosynthesis per photon) when light is spread over a larger leaf area than when it is concentrated on a smaller area. So solar tracking tends to decrease efficiency, as just discussed. Among nontracking leaves, those that are more vertical spread light over a larger area and therefore use it more efficiently, assuming that the sun is high in the sky. On the other hand, horizontal or solar-tracking leaves shade weeds more.

Similarly, taller plants may shade weeds more, but the resources they allocate to stem growth leave less resources for seed production. We might also expect crop plants that shade weeds more to shade each other more, reducing their collective yield. Based on these and similar arguments, I have long assumed that there is an inevitable tradeoff between the yield potential of a crop genotype and its ability to suppress weeds.

But recent work by Jacob Weiner, an evolutionary agroecologist working in Denmark, suggests that this tradeoff may not be so inevitable. Like Colin Donald and me, Weiner starts with the assumption that past natural selection, based on individual fitness, has left many opportunities to improve the collective performance of plant communities. Donald and

I have advocated breeding for greater cooperation among plants, in ways that enhance yield potential but sacrifice competitiveness against weeds. Weiner proceeds in a different and interesting direction. He suggests that we may be able to breed wheat plants that, in effect, cooperate to suppress weeds.[243]

Like most major agricultural improvements, Weiner's approach requires changes in crop management as well as crop genetics. Donald's wheat ideotype needs closer spacing to achieve its potential, mainly because the fewer branches and more-vertical leaves he advocates would otherwise leave gaps between plants. Solar radiation reaching those gaps can't contribute to photosynthesis, except perhaps to the photosynthesis of weeds growing in gaps between the crop plants.

Like Donald, Weiner calls for closer spacing between plants, but he also advocates a more-uniform spatial pattern. Like many crops, wheat is usually grown in rows, with closer spacing between plants in a row than between rows. With this pattern, wheat plants start to compete with neighboring wheat plants in the same row before they compete with the more-distant weeds halfway between the rows.

By increasing spacing within rows and decreasing the distance between rows, this timing could be reversed. Most weeds would face competition from a wheat plant early, and wheat plants wouldn't compete with each other until later.

Weiner notes that, when small, wheat seedlings grow faster than weed seedlings. This is mainly because resources stored in wheat's larger seeds can subsidize early growth. (See the discussion of similar subsidies in perennials in chapter 7.) Small-seeded weeds, meanwhile, must rely only on their current photosynthesis. As Darwin put it, "the chief use of the nutriment in a seed is to favor the growth of the seedlings, while struggling with other plants growing vigorously all around."[26]

So Weiner's more-uniform pattern would have wheat plants interacting with weeds early, when the wheat has a competitive advantage, while delaying competition with other wheat plants.[243] This should suppress weeds more than less-uniform seeding patterns do, while reducing harmful competition among wheat plants. Consistent with this idea, he and his colleagues have shown that suppression of weeds, even by existing wheat varieties, increases greatly when the wheat is sown at high density in a uniform pattern.[244]

Even better weed suppression may be possible if new wheat varieties are developed that take advantage of this altered pattern. The crowding responses that plants inherited from past natural selection may not be optimal from a whole-crop perspective, especially with altered spatial patterns. Growing taller and thinner to get above neighbors, when they are present (the shade avoidance response), was presumably favored by natural selection. But Weiner suggests that this trait may not be advantageous for competition with weeds.[243]

Weiner called my attention to a paper showing that crowding-induced elongation of corn seedlings has a cost. Faster elongation consumes resources that would otherwise be used for production of new leaves. The height advantage can be short-lived, while the reduction in leaf area can decrease both photosynthesis and shading of weeds.[245] So a genotype with reduced tendency to elongate in response to neighbors might actually suppress weeds better, at least with high-density uniform seeding. So long as a crop is taller than the weeds, additional height may not help.

In the previous section, I suggested that the persistence of solar tracking in high-density alfalfa stands (where it apparently reduces photosynthesis) may be maintained by natural selection because it suppresses competitors. If so, why hasn't wheat also evolved to suppress competitors?

One important difference between alfalfa and wheat is that perennial alfalfa plants can live for years, whereas interactions between annual wheat plants and weeds will last a few months, at most. It may not be worth sacrificing current photosynthesis or growth to suppress a short-term competitor. If, by tracking the sun, an alfalfa plant shades nearby weeds more, those weeds will make fewer seeds. Similarly, a competing neighbor that is shaded this year may grow more slowly, and may not survive the winter. Either way, an individual alfalfa plant that tracks the sun this year may benefit from less competition next year, from weeds or other alfalfa plants.

The situation is different for annual plants like wheat. Next year, when weed seeds produced this year grow into plants, this year's wheat plants will be long dead. Only a fraction, at most, of wheat plants growing in their spot will be their own offspring. (None will be, if farmers use purchased seed.) Therefore, natural selection among wheat plants has favored suppression of neighbors only as a side-effect of activities that directly benefit individual wheat plants. This may have left plenty of room for improvement through plant breeding.

## Biotechnology That Considers Tradeoffs: Plant Crowding Responses

We might want to modify wheat to reduce elongation in response to crowding. We might want to modify crowding responses of alfalfa to make solar tracking decrease as mutual shading increases. Fortunately, we now have a reasonably good understanding of how plants detect crowding. This information has already been used by molecular biologists to increase the yield of potatoes, as discussed shortly. But first, a little background information.

Consider a plant with two branches. If the plant has only enough resources to grow one of the two branches, which branch should grow? A branch shaded by leaves of another plant may not have much potential to photosynthesize. A branch exposed to sunlight is a better place for the plant to invest its limited resources, such as the nitrogen needed to make the photosynthetic protein, rubisco. But plants don't have brains or centralized decision-making, so where in the plant are such investment decisions made?

Somehow, shaded branches apparently monitor their own relative contribution to the success of the whole plant. If researchers apply partial shade to just one of the branches on a birch tree, that branch usually dies. But if they shade the entire tree the same amount, then none of the branches die.[246] The Australian crop scientist Victor Sadras and I have suggested that this may be because a single shaded branch sends most of its nitrogen to better-lit branches, where the nitrogen can make a greater contribution to photosynthesis.[247] Nitrogen movement wasn't measured in the birch study, but we do know that shaded leaves send nitrogen up to sunlit leaves, and then the shaded leaves often die. By committing suicide, in effect, a shaded branch may increase the representation of its genes in the next generation of birch trees, even though the flowers transmitting those genes are on other branches of the same tree. If all branches are shaded equally, however, our hypothesis predicts no net movement of nitrogen, explaining why none of the branches die when the entire tree is shaded.

But how does a branch measure its relative contribution? Maybe branches somehow measure their actual photosynthesis rate. Or they may simply detect shading by other leaves, based on how light color changes as it passes through leaves, and equate shading to low photosynthesis.

Ariel Novoplansky, an Israeli plant scientist, did a clever experiment to distinguish between these two possibilities.[248] He put different transparent barriers on two sides of a seedling. One barrier transmitted only 5 percent of the light used by photosynthesis, but didn't change the color of the light. The other barrier transmitted more light, but changed the color of the light, as if it had gone through a leaf. The seedling grew toward the barrier that transmitted less light, rather than toward the barrier that changed light color.

Under these experimental conditions, therefore, the plant grew in a direction where it would have lower photosynthesis. But basing growth direction on light color would usually have benefited the plant's ancestors. Growing toward a rock that causes color-neutral shading is better than growing toward a plant that changes light color. A plant that is small now is likely to grow, while the rock will stay the same size.

When Novoplansky visited UC Davis, he showed me a picture of a related experiment, where a tiny piece of green plastic was enough to trigger neighbor-avoidance behavior, making plants grow (as if fleeing in terror) in the opposite direction. These and similar experiments have shown that plants use light color (a low red/far-red ratio) to detect their neighbors.

Researchers in the laboratory of Jorge Casal, in Argentina, have used this response to light color to increase the yield of potatoes.[249] This is a good example of the contributions that biotechnology can make when we pay attention to tradeoffs. Casal's group focused on the same tradeoff that was key to Green Revolution increases in yield potential—namely, the tradeoff between individual competitiveness and the collective performance of plant communities.

Many plants have a similar response when a low red/far-red ratio indicates the presence of other plants nearby: they put more resources into growing taller stems, to get above their neighbors. This shade-avoidance response is the same one that Jacob Weiner suggests we might want to reduce in wheat. Increased stem growth leaves fewer resources for other purposes, such as growth of grain, weed-shading leaves, or potato tubers.

In one experiment, Casal's group grew sunflower plants far apart, so that there was little mutual shading. But they put color filters around stems, decreasing the red/far-red ratio, as if the stems were being shaded by leaves of neighboring plants. This misleading environmental cue increased stem growth, at the expense of seed yield.[250]

Although the costs of increased stem growth by sunflower plants resulted in less seed production in this experiment, it is easy to understand why this response evolved. Over most of the evolutionary history of sunflowers, a low red/far-red ratio was a reliable cue for the presence of other plants nearby. Failure to invest more in stem growth, under those conditions, meant being overtopped by neighbors. The resulting shading would have decreased an individual plant's photosynthesis, reducing resources for seed production even more than "wasteful" stem growth would have.

Casal and colleagues then moved on to potatoes. They reasoned that if they could prevent potato plants from detecting their neighbors, the plants would put more resources into growth of their edible tubers, rather than wasting resources on excessive stem growth. They transferred a light-detecting gene from another species into their potato genome, which reduced the response of potato plants to red/far-red crowding cues. Sure enough, the transgenic potato plants increased stem growth less in response to crowding and, at close spacing, they had higher tuber yield.[249]

There was apparently more than one reason for this yield increase, however, perhaps involving additional tradeoffs beyond competitiveness-versus-community. The transgenic plants had higher photosynthesis rates, apparently because the nontransgenic plants were more conservative in water use by leaves.

Why did the nontransgenic plants use less water than needed for maximum photosynthesis? It may seem like a good idea to save some water for later, rather than using it up early in the season, but natural selection will favor this trait only if the individual saving the water gets to use it later. I would have thought that transgenic plants, being less able to detect their neighbors, might have been more likely to conserve water. But apparently not. Maybe, for an isolated plant (or one unable to detect neighbors), the main "competitor" for soil water is evaporation from the soil surface, rather than use by other plants. As we scientists like to say, more research is needed.

## Increasing Cooperation among Peas in a Pod

Here is one more way that increased cooperation among crop plants might increase yield. This idea comes from Victor Sadras.[247] Consider two seeds on the same plant. Seeds are plants waiting to grow. A pollen grain has only one parent, like a sperm, but a seed has two, assuming cross-pollination. Two seeds on the same cross-pollinated plant are usually half-siblings. You might expect kin selection (based on beneficial interactions among relatives, discussed further in the next chapter; see glossary) to maintain some level of cooperation among peas in a pod. But a seed is more closely related to itself than to a sibling, especially a half-sibling. So seeds will take as many resources from their mother plant as they can get, to grow as big as possible in preparation for future competition among seedlings. A seed's chances of surviving to reproduce increase with resources per seed, but this relationship shows the usual pattern of diminishing returns—that is, decreasing marginal benefit. The maternal plant is equally related to all of its seeds. Therefore, maternal fitness is maximized by allocating resources equally among all seeds. (Some bet-hedging strategies may be an exception to this rule.[251-253])

Although maternal and paternal genes both benefit from the survival and reproduction of a seed, the paternal genes in one seed don't necessarily benefit from survival of other seeds on the same maternal plant. If anything, they are competing for the same resources. So there's a conflict between maternal plant and each cross-pollinated seed over resource allocation among seeds.

Sadras suggested that this conflict has led to the evolution of maternal mechanisms to enforce equal division of resources among seeds, including walls dividing pods and certain features of the seed coat. These may limit the ability of any one seed to grab more than its share of resources. But, he suggested, these mechanisms also limit the maximum potential growth of seeds, under favorable conditions. This tradeoff was apparently acceptable to natural selection, perhaps because conflict among seeds was universal, whereas favorable conditions were rare.

Today, however, the opposite may be true. In agriculture, conditions may often be favorable enough that seeds could grow significantly larger, were it not for maternal controls. Many important crops are now self-pollinating, but they may retain maternal resource-allocation controls from cross-pollinated ancestors. Because self-pollinating crops have no intrinsic conflict between maternal and paternal genomes, we may be able to dispense with some of these maternal controls. This would potentially increase yields in good years, without any significant cost in poor years.

## Cooperation among Chickens

Like plants, farm animals may have inherited some unfortunate traits from ancestors involved in past evolutionary arms races. Can we increase their productivity by selecting for greater cooperation, reversing past natural selection?

Some evolutionary biologists—but not those I admire most[254]—have objected to Richard Dawkins's use of metaphors, like "selfish" genes[142] or "arms races"[230] to explain evolutionary concepts. However, I found his books quite useful when I first started exploring these topics. So I looked forward to talking to him, a few years ago, after he spoke at a small college nearby. After I outlined my ideas for this book, he called my attention to some very interesting experiments that William Muir has done with chickens.[255]

Farmers and poultry breeders have been selecting for increased egg production for centuries, by killing and eating the hens that lay the fewest eggs, then raising successive generations preferentially from hens that produce the most eggs. Even before domestication, the wild ancestors of chickens must have been under selection for increased egg production, despite tradeoffs between size and number of eggs and between current and future reproduction.[256, 257] Given all that past selection, you might think there would be little opportunity for us to select for still greater egg production today.

But that pessimistic view assumes that current tradeoffs are similar to those that constrained past evolution. Muir worked with hens raised in groups of four, as they sometimes are in commercial egg production. He based his selection scheme on the hypothesis that there can be tradeoffs between individual egg production and egg production by the whole group.

Over much of their evolutionary history, hens have competed among themselves. Those who were more aggressive tended to get more than their share of food, allowing them to lay more eggs and have more descendants. Although success in fighting increased the relative egg production of the most-aggressive hens, it decreased egg production by the losers even more, thereby reducing total egg production.

So Muir took a novel approach to selecting for increased egg production. Rather than breeding from the individual hens that laid the most eggs, he bred from the groups—four hens per group, raised together—that laid the most eggs. The most-productive groups turned out to be those where hens had less genetic propensity to peck each other. In only six generations of selection, there was a major decrease in fighting-related injuries and a 30 percent increase in the percent of hens laying an egg each day, from 52 percent of hens to 68 percent.[255]

Selection based on group-level traits, like egg production per group, is known as *group selection*. Group selection deliberately imposed by humans (as in the chicken example just discussed) can have major evolutionary effects. But most evolutionary biologists agree that, except when groups consist mainly of close relatives, differences in the survival of groups have much less evolutionary effect than differences in survival and reproduction among individuals within groups, at least in nature. So if we want group-level benefits that conflict with individual fitness, we will have to impose group selection ourselves.

Nature films often miss opportunities to point out differences between group selection and kin selection, which is based on benefits to relatives. For example, one program on fresh-water ecosystems, narrated by nature expert David Attenborough, shows a small family of otters working together to chase a crocodile away. Later in the same program, however, hundreds of much larger (but unrelated) wildebeest failed to work together while crossing a river. So they fell prey, individually, to crocodiles.

As usual, a selfish-gene perspective helps explain the difference.[142] Alleles leading to behavior that increases the survival of others that are particularly likely to have the same alleles (as in the otter family) can increase in frequency through kin selection, even if that behavior has individual risks. A simple explanation of the math behind kin selection is given in the next chapter.

Working together, wildebeest could presumably put up a stronger group defense than the much smaller otters. But each wildebeest that dares to confront a crocodile would risk injury, even if the crocodile is driven off, and the beneficiaries are not particularly likely to share the same alleles for cooperation. So individual selection undermines cooperation.

When group selection and individual selection act in opposite directions like this, which is more powerful? Wildebeest herds that have mostly cooperative members would suffer less predation and reproduce more, increasing the frequency of alleles for cooperation in the wildebeest population.

But the larger the herd size, the less the chance that random assortment of unrelated individuals among herds will lead to significant differences in overall cooperation alleles among herds. Without such differences among groups, group selection is ineffective. Also, what if cowardly, uncooperative wildebeest whose herd suffered, because they didn't help defend it, then switched to a herd defended by more-heroic wildebeest? By doing so, the cowardly individuals would enhance their chances of surviving and reproducing, thereby increasing the frequency of alleles for cowardly defection from group defense. So group selection should be most powerful when group size is small and when there is less migration between groups.

Mathematical modeling suggests that a group of fewer than 25 animals, with less than 5 percent migration between groups per generation, is needed for group selection to counter individual selection.[138, 139] Muir's experiments with chickens had no migration between cages, and only four hens per cage, so it is not surprising that there was a strong response to selection.

One published experiment seems to show, surprisingly, that group selection by humans can sometimes be successful, even with many individuals per group. William Swenson and colleagues grew plants in pots containing soil with diverse microbial communities. In each cycle of selection, they used soil from the three pots with the best plant growth (of 15) to inoculate the next set of 15 pots.[258] (Some orchard managers once followed a similar practice, supposedly, inoculating newly planted orchards with soil from the healthiest old orchards.) Another set of plants was inoculated each cycle with soil from plants that grew least.

Plants varied in growth over the first ten cycles, but differences among treatments were small and growth was sometimes better with soil microbes selected for worse plant growth. In cycle 10, both treatments had very poor plant growth, worse than anything seen previously. But suddenly, in cycle 11, plants with microbes selected for more plant growth grew several-fold better than those with microbes selected for poor plant growth. This difference persisted until cycle 15, when most of the plants in both treatments were killed by a fungal pathogen.

I would like to see this experiment repeated by another lab before drawing firm conclusions, and it would be interesting to know which microbes were responsible for the differences. If these results were confirmed, that would suggest that human-imposed group selection on microbial populations can sometimes work, even with large group sizes. One possible explanation is that the evolutionarily effective population size of a large group of bacteria that are almost identical genetically (because, when bacteria divide, they make two nearly exact copies of themselves) can be much less than the number of cells.

In any case, the apparent success of this experiment and that by Muir both confirm the near-consensus among evolutionary biologists that group selection is rare in natural ecosystems. This was not the conclusion drawn by Swenson's group. But if group selection for beneficial plant-associated microbes had already been common in nature, that would have left little opportunity for further improvement by Swenson and colleagues. Similarly, if group selection had already optimized the collective behavior of hens, there would have been little room for improvement by Muir.

The limited evolutionary role for selection among groups of nonrelatives, in nature, has been supported by a recent mathematical analysis that "failed to find any justification for group adaptationism for scenarios in which within-group selection is permitted."[136] In other words, even a little individual selection can undermine group selection, if they act in opposite directions.

But if group selection has been rare in nature, that is actually good news for agriculture. It means that there may still be many opportunities for human-imposed group selection to improve the collective performance of groups of farm animals or crop plants.

## Group Selection for Crop Genetic Improvement

Plant breeders routinely use group selection, although they don't usually call it that. At some point in the breeding process, they grow plants in groups (usually consisting of a single genotype) and choose those that collectively produce more grain or other useful product. Group selection for yield may explain decreases in tassel size and more-vertical leaves of corn over decades, which "did not result from direct selection efforts of the breeders."[84]

This form of group selection has some limitations, however. With small plots separated by unplanted alleys, edge effects can result in unrealistically high yields. Leaving out the alleys can be even worse. Then, competition for light or other resources among neighboring genotypes can lead to incorrect estimates even of relative performance. For example, taller genotypes next to shorter genotypes may be incorrectly identified as having greater yield potential.[259, 260]

Using larger plots and measuring yield only in the center of each plot is a widely recognized method of avoiding edge effects. But giving each genotype larger plots (usually several replicates scattered over a field, to correct for variability in soil quality) means fewer genotypes can be tested in a field of a given size. Nonetheless, some breeders manage to test hundreds of genotypes.

Advances in automation may make it possible to practice group selection on an even larger scale. *Yield monitors* built into harvesting equipment can measure grain yield from each small area of a field as it is harvested. If seeds were planted so that identical or similar genotypes grew together, a yield monitor could be modified to separate out and save seeds from parts of a field that are particularly high yielding, perhaps because they have more-cooperative genotypes. We would probably need to correct for past yield in the same part of the field, a process that could be computerized. A few cycles of such group selection would be enough to see whether this approach would be effective.

Group selection could be based on criteria other than yield. For example, semiautomated, human-mediated group selection could be used to identify crop genotypes that cooperate to use water more efficiently. Infrared thermometers that monitor leaf cooling by transpiration have already been used to identify wheat cultivars that use more water. Varieties that use more water tend to have higher yield, as one might expect.[261] But, in combination with a yield monitor, one could also select for cultivars with high yield *relative to their water use*—that is, for high water-use efficiency.

## Perspective

Past evolutionary arms races have resulted in plants—and chickens!—that compete vigorously with their neighbors for resources, even when that competition reduces their collective productivity. In some cases, we have reversed past natural selection for competitiveness, often inadvertently. The resulting genotypes are more productive because they are more cooperative, in the sense that they suppress each other's growth less. In my opinion, further exploration of this tradeoff-based approach is the most-likely route to increasing yield potential.

One practical tradeoff is that crop genotypes that suppress each other less may also suppress weed growth less. To the extent that weeds can be controlled in other ways, this approach probably still has considerable untapped potential. However, I look forward to the results of Jacob Weiner's attempts to make wheat suppress weeds in ways that are less likely to suppress fellow wheat plants.[243]

Colin Donald identified several traits that can improve the collective performance of crops like wheat, at some cost to individual competitiveness.[112] Drysdale wheat, which performs well under both wet and dry conditions,[9] may also have given up some competitiveness, if it turns out to photosynthesize less at times of day when water is used least efficiently.

Further exploitation of individual-versus-community tradeoffs could be based on analysis of specific traits, like solar tracking, using experiments and computer models. It may also be possible to select for more cooperative crops and livestock without identifying the key traits in advance, through automated human-mediated group selection.

Achieving optimum levels of within-species cooperation (for example, among wheat plants or chickens) may often require human intervention. But there is one form of within-species cooperation that is often seen in wild animals, and maybe sometimes in wild plants: cooperation among close relatives, due to kin selection. In the next chapter, where I discuss cooperation between two species, I will explain how cooperation among related individuals of one species can also help maintain cooperation between species.

# 9

## What Could Work Better

COOPERATION BETWEEN TWO SPECIES

CROP PLANTS OFTEN DEPEND on other species for pollination, nutrients, or defense against pests. In this chapter, I discuss some examples of between-species cooperation, the evolutionary tradeoffs that can undermine such cooperation, and opportunities for improvement.

## Cooperation between Two Species Is Common, but So Is "Cheating"

Cooperation between species might seem even more difficult than cooperation among wheat plants or among chickens, but the opposite may be true. Plants compete with neighboring plants for sunlight, water, and soil nutrients. A plant that takes up more nitrogen than it needs might thereby suppress the growth of a neighbor, which would otherwise compete with it for water. As discussed in the previous chapter, a plant might benefit from shading a neighbor, even if that requires solar tracking, which slightly reduces photosynthesis of its own lower leaves. More generally, cooperation among individuals may not increase individual fitness, if they all depend on a limited pool of the same resources.

On the other hand, cooperation between species that use *different* resources is less likely to be undermined by competition between them. For example, plants and bees don't compete with each other for sunlight or soil nitrogen. Although both species need water, they don't compete for it. During drought, plants get most of their water from roots deep in the soil, beyond the reach of bees. So while a plant may benefit from the death of a neighboring plant, bees are more valuable to plants alive than dead.

Therefore, it is not surprising that we have many more examples of cooperation between species than between unrelated individuals of the same species. Even between-species cooperation can be undermined by evolutionary tradeoffs, however. These tradeoffs usually involve conflicts of interest between partners, even though they also have shared interests. *Cheating*, where individuals exploit cooperative partners, is widespread, as discussed shortly.

Like tradeoffs linked to within-species competition, tradeoffs responsible for between-species cheating can represent opportunities for humans to improve on nature. With *mutualisms* (mutually beneficial interactions between species), however, we are working to enhance mostly positive interactions, perhaps accelerating natural selection rather than reversing it. Let's start with some examples of between-species cooperation, in nature and in agriculture, before discussing conflicts of interest and opportunities for improvement.

Most wild plants rely on mutualistic symbiosis with mycorrhizal fungi for phosphorus and sometimes for other benefits. Plants in the legume family (bean, clover, and so on) often host bacteria known as rhizobia. Inside legume root nodules, mutualistic rhizobia capture nitrogen gas diffusing in from air spaces in the soil and convert it to forms the plant can use to make protein. In return, rhizobia and mycorrhizal fungi get most of their energy from their plant hosts.[262]

Our most-important grain crops pollinate themselves or rely on wind-borne pollen, but many other crops depend on insects for pollination. Animals that eat fruits and disperse seeds in their droppings increase the fitness of wild plants, although we usually prefer to handle seed distribution of crops ourselves.

Some of the most interesting collaborations between species involve protection. Fungal endophytes (see glossary) live inside plants without causing obvious disease. Endophytic fungi living inside leaves of the cacao tree help protect this crop (the source of chocolate) from harmful fungi.[263] Endophytic fungi in some grasses produce toxins that can discourage animals from eating the plants.[264] Unfortunately for agriculture, these animals can include cattle, whose growth rate can be decreased almost 50 percent by high levels of endophytic fungi.[265]

Acacia trees feed and house ants, which vigorously defend their hosts from other insects and from larger animals. Ants that nest in a particular rain-forest tree even kill competing trees, creating single-species (*Duroia hirsuta*) plant monocultures known as "devil's gardens." The ants kill other plant species by injecting them with formic acid, a natural toxin they also use against insect prey.[266]

All of these interactions are rightly considered mutualisms, typically providing net benefits to both species. But mutual benefit does not exclude the possibility of conflict. Cooperative interactions can be exploited, either by the usual partners, or by other species. In the next section, I will give some examples of such conflicts. Then, I'll discuss how conflict evolves and how we may be able to improve two-species partnerships useful in agriculture.

## Conflicts between Partners

Let's start with rhizobia, best known for providing their host plants with nitrogen. There are only a few reports of rhizobia that actually reduce plant growth.[267] However, strains that provide little or no net benefit are common enough in some soils to be a significant problem worldwide.[268–272] Even rhizobia that provide their host with some nitrogen can be harmful, by outcompeting better strains for places in root nodules. Should we call these strains "cheaters"?[273]

Mycorrhizal fungi also vary in the net benefits they provide to their host plants.[274] Benefits depend on host and fungal genotype and on soil conditions. If there is plenty of soil phosphorus available to plant roots, the cost of supporting the fungus may exceed its benefits.[275]

Pollination of plants by insects can also be undermined by cheating. Robber bees cut into flowers to take nectar without transporting pollen,[276] sometimes chasing away legitimate pollinators. The larvae of insects that pollinate figs and yuccas also eat some of the seeds inside the developing fruit.[277] Some moths descended from yucca pollinators eat seeds without pollinating at all.[278] Fruit-eating animals that digest seeds, rather than dispersing them, are a problem for many plants. Chili peppers have evolved a sophisticated solution: their spicy taste discourages seed-digesting mammals, but not seed-dispersing birds, whose mouths lack capsaicin receptors.[279]

Apparent alliances based on protection are not always what they seem. Endophytic fungi in some wild grasses may provide little protection, in contrast to those in cultivated pasture grasses.[280] One tree species protected by ants is frequently castrated by its own guards; they remove flowers, which don't benefit the ants, while protecting leaves, which do.[281] Some ants hosted by acacia trees have been found to prune branches as well as flowers, apparently to keep the tree from touching neighbors occupied by more-aggressive ants.[282]

By creating "devil's gardens" dominated by *Duroia hirsuta*, ants create a local monoculture that attracts insects that eat leaves of this host species. Insect damage to *Duroia hirsuta* is higher in the gardens than in parts of the forest with more tree diversity. The insect damage was even worse if ants were excluded from the monocultures they themselves created, however, so the ants do apparently provide some protection from insects as well as from competing trees.[266]

With all this conflict, why doesn't cooperation between partners break down? Some mutualistic symbioses apparently *have* broken down. For example, some free-living fungi are descended from ancestors that spent most of their lives in symbiosis. Some were previously in mycorrhizal association with plants.[283] Others are descended from fungi that partnered with algae to form lichens.[284] Often, however, cooperation persists despite cheating by one or both partners.[285] Why do they stay together?

## Symbiotic Nitrogen Fixation and the Tragedy of the Commons

Basically, mutualistic partners stay together because of the children. I am referring, of course, to cooperation's reproductive benefits for each partner. Unlike partners of the same species, those of different species don't have any shared offspring.

Consider rhizobia inside a legume plant's root nodules. Each of these bacterial cells receives resources from the plant. A rhizobial cell could use those resources for its own immediate reproduction, or it could use them to power nitrogen fixation, taking up nitrogen from the atmosphere and giving it to the plant host. So rhizobia face tradeoffs in the allocation of resources, just as plants do.

Even though rhizobia give their host almost all of the nitrogen they fix, rhizobial investment in nitrogen fixation isn't pure altruism. With more nitrogen, the plant will grow more, photosynthesize more,[286] and perhaps supply more photosynthate to the rhizobia. So the rhizobia in each plant's root nodules benefit, collectively, if they all invest in the nitrogen fixation that helps their shared host grow. This might seem to favor cooperation among rhizobia infecting a given plant.

Yet cooperation among rhizobia is a problem. The opportunity cost of investing more resources in fixing nitrogen is paid by individual cells. If benefits from increased plant growth are shared by all the rhizobia in a given plant's root nodules, will natural selection favor rhizobia that invest in nitrogen fixation?

Maybe not. After all, the failure of one cell among millions to fix nitrogen won't have much effect on plant growth. Yet that one cell could use the resources it saved for its own reproduction. Furthermore, the other rhizobia on the same plant are potential competitors for future plant hosts, so helping the plant indirectly helps competitors. The situation appears to be a *tragedy of the commons*.

That's the term biologist Garret Hardin used to describe the dilemma posed by shared resources, such as grazing land or ocean fisheries.[287] If just one person puts twice as many cows on the common grazing land, she will have twice as much milk to sell, and the land can probably support one more cow without damage. But if everyone follows that same logic, they will double the total cow population, causing such severe overgrazing that milk production decreases. Similarly, if one fisherman catches twice as many fish, that may increase his income without hurting fish populations, but if everyone catches twice as many fish, the fish may go extinct. Hardin suggested that many other human activities, from air pollution to human reproduction, can involve tragedies of the commons. This is because some of the costs of activities that benefit individuals are shared by a larger community.

As I was working on this chapter, political economist Elinor Ostrom won the Nobel Prize, partly for research showing that traditional human communities often manage this sort of problem rather well.[288] Tragedies of the commons do seem to be abundant in the natural world, however. Plants overinvesting in stem growth, aggressive root exploration of soil beneath neighboring plants, and wasteful solar tracking are examples from the previous chapter.

What about rhizobia? A key point is that rhizobia, like other bacteria, duplicate their entire genome when they reproduce. Human children inherit about half of their genes from each parent, but bacteria inherit their one parent's entire genotype. Rhizobial reproduction therefore generates clonal populations that are genetically identical, or nearly so. If the rhizobia inside a nodule are all descended from one individual cell, then each rhizobium is surrounded by millions of genetically identical clonemates. This genetic similarity can make evolution of cooperation much more likely.

## Kin Selection and Within-species Cooperation

This principle doesn't apply only to bacteria. In general, within-species cooperation evolves more easily among kin, because they share more alleles than nonrelatives. Some might argue about whether members of a rhizobial clone should be called kin, but they do share essentially all of their alleles. An allele whose effects preferentially increase the survival or reproduction of other individuals with that same allele will become more common over generations, even if some of those alleles are in brothers, nieces, or clonemates.

*Preferentially* is the key word here. Indiscriminate aid to those with competing alleles is an admirable human trait, but it is not usually favored by selection, so it is rare in species other than humans. Natural selection based on shared alleles is often known as *kin selection*, a key evolutionary concept pioneered by evolutionary biologist William Hamilton.[140]

Hamilton showed that an allele for sacrificing some of one's own fitness (potential reproduction), while increasing the fitness of another individual, can sometimes increase over generations. This will happen, he showed, if and only if the "altruist's" fitness cost $c$ is less than the fitness benefit $b$ to the recipient, multiplied by a term $r$, which Hamilton called "relatedness." That is, an altruism allele will spread if

$$c < b\,r$$

The $r$ term can correspond to the degree of relatedness within a family, but I prefer to call it "Hamilton's $r$," for reasons that will become clear.

Hamilton's rule is one of the most important equations in all of biology, yet it is widely misunderstood. It seems to predict that species would evolve a tendency to help even quite distant relatives, perhaps any member of their species, if the cost were low enough. For example, I have seen Hamilton's rule used to explain generosity to beggars. Even if two people are very distantly related, the argument went, the fitness cost $c$ of a dollar to a rich person is tiny relative to the fitness benefit $b$ to a starving person.[34]

Unfortunately for beggars, Hamilton's $r$ is more accurately seen as how much *more* closely two individuals are related to each other than either is to the overall population,[289] as explained shortly. For two random people, therefore, Hamilton's $r$ is as likely to be slightly negative as slightly positive.

The evolutionary consequences of Hamilton's rule do not require intelligence, although the ability to recognize relatives can be helpful, so it can be applied to cooperation among plants or bacteria. For example, it has been claimed that taller trees subsidize their shorter neighbors by transferring photosynthate through mycorrhizal fungi that are connected to both plants.[290] Why might a tree share photosynthate with seedlings?

A tall tree, with good access to sunlight, may lose little fitness when it shares some photosynthate, relative to the fitness benefits gained by shaded seedlings. In other words, the benefit to the recipient seedling can be much greater than the cost to the donor tree—that is, $b >> c$. Perhaps many of those seedlings are its own offspring, sharing half of the taller tree's genes. If sharing half of one's genes always results in a reasonably large Hamilton's $r$ (0.5, say), that would seem to predict sharing, even if some unrelated seedlings also benefit.

Understanding Hamilton's $r$ is so central to understanding cooperation, yet so widely misunderstood, that I need to discuss it at length. Fortunately, there is an easy way to visualize Hamilton's $r$, without advanced math. This approach was developed by Oxford professor Alan Grafen.[291]

There are two basic ways in which individuals might preferentially benefit others likely to share their alleles. Many animals can recognize their own relatives and direct aid preferentially to them. Bacteria can't usually recognize individuals, or so we assume. But because they reproduce by copying their entire genomes and then dividing, they often occur

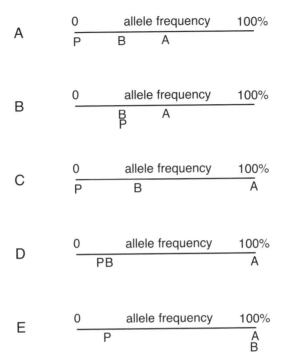

Figure 9.1. Grafen's visualization of Hamilton's *r*, which predicts evolutionary trends in the frequency of alleles for cooperation, as explained in text.

in clumps of genetically identical clonemates. Therefore, a bacterium that pays some cost to benefit all the bacteria in its immediate neighborhood may benefit thousands or millions of cells that share nearly all of its alleles. In such cases, the *b* in Hamilton's rule will represent a total summed over many beneficiaries. I will discuss kin selection in animals first, followed by the analogous process in bacteria.

Grafen focuses on an individual with an allele for *altruism* (benefiting others at some cost to itself) and asks how its altruistic actions would affect changes in the frequency of the altruism allele in the local population. He plots the frequency of the altruism allele (ranging from 0 to 100 percent) in the evolving population (*P*), the frequency in the subset of the population that benefits from the focal individual's altruism (*B*), and the frequency in the focal altruist itself (*A*), as shown in figure 9.1 for various situations.

Let's start by assuming that the altruism allele is rare, so that its frequency in the population, *P*, is essentially zero (figure 9.1A). Animals (and cross-pollinated plants) have two alleles of each gene, one from each parent. Since we have assumed that the altruism allele is rare, our focal individual probably has only one copy, inherited from its mother or father, but not both. So the frequency of the altruism allele in the focal individual, $A$, is 50 percent, as plotted in figure 9.1A.

Now we need to consider possible beneficiaries of our focal individual's altruism. Assume, for now, that those beneficiaries are the individual's children. The average frequency of the altruism allele in that pool of potential beneficiaries is plotted as 25 percent in figure 9.1A.

Why is 25 percent a reasonable estimate of the frequency of the altruism allele in our focal individual's children? We have assumed that the overall population, including our focal individual's potential mates, has an altruism-allele frequency near zero. So our focal individual's children won't get a copy of the altruism allele from that other parent. Half of the offspring will get one copy from our focal individual, giving them the same 50 percent allele frequency (one allele of two) as the focal individual, while half of them won't get even one copy, giving them an altruism-allele frequency of zero. So the average altruism-allele frequency in the focal individual's offspring is (50 percent + 0 percent) / 2 = 25 percent. I have therefore plotted $B$, the altruism-allele frequency in the beneficiary pool consisting of our focal individual's own offspring, as 25 percent.

Under our assumptions, our focal individual's siblings also have an average altruism-allele frequency of 25 percent. Because the allele is rare, most individuals don't have it at all, and those that do will have only one copy. (I'm assuming that inbreeding is rare.) One of our focal individual's parents must have had the altruism allele, but presumably just one copy. So half of that parent's children (our focal individual's siblings) will get one copy and half will get none, giving them the same average altruism-allele frequency as the focal individual's own children.

Potential beneficiaries who are more distantly related to our focal individual (cousins, say) will have lower values of $r$. Before either Hamilton or Grafen published their analyses, evolutionary biologist J.B.S. Haldane supposedly joked that he would die for two brothers or eight cousins. What definition of Hamilton's $r$ would make Haldane's statement consistent with Hamilton's rule?

If our focal individual reproduces, that will tend to increase the frequency of its alleles in the overall population. For example, its reproduction will increase the frequency of the altruism allele, because (in figure 9.1A) $A$ is greater than $P$. Alternatively, the focal individual could forgo reproduction in order to help one or more siblings reproduce. Reproduction by one sibling would also increase $P$, but only by half as much as if the focal individual had reproduced, on average. Why? Because the frequency of the altruism allele in the average sibling is only 25 percent, versus 50 percent in the focal individual.

So the focal individual would have to help two siblings reproduce to have the same evolutionary effect as reproducing himself. (Similarly, he would have to save two brothers from drowning to make it worth drowning himself.) Paying a cost of one offspring to help two siblings reproduce is consistent with Hamilton's rule (under our assumptions of figure 9.1A) only if we set Hamilton's $r = 0.5$ when the beneficiary is a sibling.

So far, Hamilton's $r$ seems to be equivalent to our usual genealogical definition of relatedness: 0.5 for siblings, 0.125 for cousins, and so on. Thus, we might expect a tree to share photosynthate with one of its own seedlings, if doing so increased the seedling's reproduction twice as much as it reduced the tree's own reproduction. This is the usual understanding of kin selection among biologists. But we need to dig a little deeper.

What if the entire population consists of the focal individual's own offspring? (Such extreme examples often make complex points easier to understand.) In that case, the potential beneficiary population is exactly the same as the overall population, so $P = B = 25$ percent, as drawn in figure 9.1B. So reproduction by our focal individual's offspring, the potential beneficiaries of its altruism, will have *no effect* on the frequency of the altruism allele in the population. If helping potential beneficiaries would not actually increase the frequency of the altruism allele, then a correct formulation of Hamilton's rule must predict that altruistic helping will not evolve, however low the fitness cost of helping, and however large the fitness benefit to the offspring.

What value of Hamilton's $r$ will make this true? Hamilton's $r$ must be zero. Yet the genetic relatedness of offspring in figure 9.1B is the same as in figure 9.1A—namely, 0.5. We must therefore conclude that Hamilton's $r$ isn't always equal to our usual definition of genealogical relatedness. So how *should* we define Hamilton's $r$?

Grafen solved this dilemma by defining Hamilton's $r$ as *how far B is toward A from P*.[291] That would make Hamilton's $r = 0.5$, in figure 9.1A, but $r = 0$ in figure 9.1B. This is exactly what we need for Hamilton's rule to work.

If the overall population has essentially zero relatedness to our focal altruist, as in figure 9.1A, then Hamilton's $r$ will correspond to our usual definition of relatedness. But often, as in figure 9.1B, Hamilton's $r$ will be less than we would expect from genealogical relatedness alone, making evolution of altruism less likely.

This may explain why, as it turns out, trees don't really share photosynthate with seedlings. Trees do supply mycorrhizal fungi with photosynthate, in exchange for phosphorus, but the fungi don't usually pass it on to other trees. Carbon from "donor" trees that seemed to show up in roots of recipient seedlings actually stayed in fungi attached to the roots of those seedlings.[292] If a tree is surrounded by its own seedlings, there is no reason to give one offspring resources that will help it outcompete another offspring.

## Microbial Analogs of Kin Selection and Rhizobial Mutualism

Hamilton's rule works a bit differently for bacteria, including nitrogen-fixing symbiotic rhizobia. First, bacteria typically have only a single allele of each gene, so a focal cell's altruism-allele frequency, $A$, can't be 50 percent. It's either 0 or 100 percent. Also, bacterial altruism often benefits any bacteria in an altruist's immediate neighborhood, rather than being directed toward particular individuals.

Consider a rhizobial cell with one copy of an altruism allele for fixing a lot of nitrogen, which directly benefits its legume host and indirectly benefits other rhizobia dependent on that host. The cell has zero copies of an alternative cheater allele, for diverting resources from nitrogen fixation to its own reproduction. This gives our focal rhizobium cell an altruism frequency, $A$, of 100 percent, as shown in figure 9.1C. Fixing nitrogen has some net cost, $c$, to the rhizobial altruist, and some net benefit, $b$, mediated through benefits to the host, which is shared by all the other rhizobia on the same plant.

Those rhizobia constitute a beneficiary group (with altruism-gene frequency = $B$), which includes the altruist's clonemates in the same root nodule. Those clonemates are genetically identical to our focal altruist—that is, they all have the altruism allele, because they are all descended from a single nodule-founding cell. But potential beneficiaries also include other rhizobial clones, without the altruism allele, in other nodules on the same plant. Our focal cell's nitrogen fixation, inside a nodule, has no effect on most of the local rhizobial population (with altruism-gene frequency $P$), which includes rhizobia on other plants nearby, or in the soil.

How does our focal rhizobial cell's altruistic nitrogen fixation affect the evolution of altruism? That is, how does it change $P$, the frequency of the altruism gene in the overall population? It depends on the frequency of the altruism allele in the overall population and in the beneficiary population, that is, on $P$ and $B$.

In figure 9.1C, I've assumed that about 1/3 of beneficiary rhizobia, those that benefit from our focal altruist's nitrogen fixation because they're on the same plant, have the altruism allele. Maybe it's a small plant with only six nodules (each containing 10 million rhizobia), of which two nodules contain the altruist genotype. I've also assumed that the altruist-allele frequency in the overall population is essentially zero. So in figure 9.1C, $P$ is drawn at zero, $B$ at 33 percent, and $A$ at 100 percent.

If our focal cell reproduces (by dividing into two identical cells), that will add one more altruist to the population, increasing $P$ slightly. But if it altruistically forgoes reproduction and instead fixes more nitrogen, that will slightly increase plant health and the chances of reproduction for all rhizobia in nodules on that same individual host plant. Suppose our focal cell's celibacy allows enough additional nitrogen fixation that three rhizobial cells reproduce that wouldn't have otherwise. Because 1/3 of them have the altruism gene, on average, reproduction of those three cells will add one copy of the altruism gene to the overall population, the same as if our focal altruist cell itself had reproduced. It's a toss-up.

If we use Grafen's definition of Hamilton's $r$, will Hamilton's rule correctly predict natural selection's indifference to altruism, in this situation? Yes. $B$ is 1/3 of the way toward $A$ from $P$, so Hamilton's $r = 1/3$. We have assumed that the benefit to recipients, $b$, is three times the cost $c$ to the focal altruist, so $c = 3c\ (1/3)$. So Grafen's version of Hamilton's $r$ works.

What if there are already many altruistic rhizobia in the overall soil population? This is actually likely, unless the altruism allele is a recent mutation. And what if only a small fraction of the nodules on the plant are occupied by our focal altruist's clonemates? This also seems to be the usual situation, with an average of ten or so different rhizobial strains per plant.[293] Combining these two concessions to reality gives the altruism-gene frequencies shown in figure 9.1D.

With these more realistic assumptions, helping other rhizobia on the same plant (beneficiaries whose altruism-gene frequency is $B$) has only slightly more effect on the frequency of the altruism allele as helping a random member of the population (with altruism-gene frequency $P$). In other words, indirectly helping beneficiaries whose relatedness to the focal strain is only slightly above the population average doesn't increase the overall frequency of altruists much.

Grafen's definition of Hamilton's $r$ still works in this situation. $B$ is only a small fraction of the distance toward $A$ from $P$, so Hamilton's $r$ is small, perhaps 0.1 or less. So the reproductive opportunity cost, $c$, of altruism (fixing more nitrogen) would have to be less than 10 percent of the benefit to beneficiaries for fixing nitrogen to evolve or persist.

My overall conclusion from the preceding is that we shouldn't necessarily expect individuals to help their neighbors, just because those neighbors tend to be somewhat related to them. Instead, the evolution of altruism based on relatedness will usually require some way to direct aid specifically to one's closest relatives. Humans and other animals can do this. Some plants may, too,[294] although I don't understand how. But can a rhizobial cell preferentially direct the benefits that come from fixing nitrogen to its own clonemates?

It turns out that they can, but not simply by increasing the growth and photosynthesis of their shared plant. Figure 9.1D is consistent with the large number of rhizobial strains typically seen on each plant and with those strains being a random sample of the soil population. Figure 9.1D predicts such a low value of Hamilton's *r* that we would have expected rhizobial cheaters that divert all available resources from nitrogen fixation to their own reproduction to have displaced altruistic, nitrogen-fixing strains long ago.

Yet that hasn't happened. So somehow, nitrogen fixers must benefit other nitrogen fixers more than they benefit cheaters. But how?

## Host Sanctions Maintain Microbial Cooperation with Plants

Back in 2000, my analysis of legume-rhizobium cooperation as a tragedy of the commons was less sophisticated than the preceding discussion. I merely suggested that if nitrogen fixation benefited all the rhizobia on the plant equally, by improving plant health, then a rhizobial strain that invested more of its resources in nitrogen fixation could end up benefiting competing strains on the same plant, including cheaters that invested more in their own reproduction.[295]

During a brief visit to my lab, evolutionary theorist Stuart West did a more thorough analysis. He showed mathematically that, with several rhizobial strains per plant, Hamilton's *r* is very low, consistent with figure 9.1D. So, if plants treat all rhizobia alike, strains that invest little or nothing in nitrogen fixation should have taken over long ago.[296] But, again, we know that this hasn't happened. Although rhizobial cheaters are apparently common enough to be an agricultural problem in some soils, they aren't as common as West's initial model predicted.

So he went on to model what I called the *sanctions* hypothesis.[295] Under this hypothesis, plants monitor nitrogen fixation by individual nodules and impose sanctions that reduce the fitness of rhizobia in poorly performing nodules.

If each nodule contains only one genotype of rhizobia, then the Grafen diagram for an individual root nodule (figure 9.1E) is very different from that for the plant as a whole (figure 9.1D). With host sanctions against individual nodules that fix less nitrogen, the beneficiaries of our focal cell's nitrogen fixation are mostly its own clonemates inside the same nodule. They all have the same altruism (nitrogen-fixation) allele frequency, for example, 100 percent. The benefit they all receive is holding off the host sanctions that would be triggered if they don't fix enough nitrogen. With one genotype per nodule, and benefits at the level of individual nodules, Hamilton's *r* would be close to one, unless the overall population frequency *P* is already 100 percent.

West's model showed that host sanctions against nodules that fix less nitrogen would impose strong selection for nitrogen fixation, in contrast to very weak selection from shared, whole-plant benefits. A recent mathematical analysis of plants interacting with mycorrhizal fungi came to similar conclusions.[297] But do host sanctions against nodules that fix less nitrogen actually exist? To find out, we decided to manipulate rhizobial cooperation, to see how host plants responded. Toby Kiers worked with my technician, Bob Rousseau, to grow soybeans in ways that would let us control the nitrogen gas concentration around their root nodules. When a nodule is exposed to a nitrogen-free atmosphere—with argon gas instead of nitrogen gas—the rhizobia inside stop fixing nitrogen. They compared two nodules on the same plant, randomly giving one nodule air and the other nitrogen-free air. After 10 days, they estimated the number of rhizobia in each nodule by counting bacterial colonies on culture plates.

If there were no host sanctions, then preventing rhizobia from fixing nitrogen could free up resources, which the rhizobia might use for their own reproduction. So this *no-sanctions hypothesis* would predict that, if anything, nodules in nitrogen-free air would contain more rhizobia.

But we found the opposite, disproving the no sanctions hypothesis. Soybean rhizobia prevented from fixing nitrogen reproduced less than half as much as those that had fixed nitrogen.[298]

Ryoko Oono, who earned a PhD with me more recently, found similar sanctions in alfalfa and pea, even though rhizobia in their nodules are significantly different from those in soybean.[299] In alfalfa and pea nodules, rhizobia lose the ability to reproduce, like worker bees, as they gain the ability to fix nitrogen. But the nodules also contain reproductive clonemates of the nitrogen-fixing rhizobia, analogous to queen bees, and those are affected by sanctions.

Plants can apparently impose sanctions on mycorrhizal cheaters, too. The most-convincing evidence comes from Toby Kiers and colleagues, who used a clever new technique to show that a strain that provides the host plant with more phosphorus gets more carbon in return, even if the two strains are intermingled on the same root. Plants were given a heavy isotope of carbon, which they passed on to mycorrhizal fungi. RNA from the mixed-strain fungal community was separated into heavy and light fractions and then sequenced. The heavier fraction, which had gotten more carbon from the plant, was dominated by RNA sequences corresponding to the better strains.[300]

We aren't yet sure how plants impose sanctions, especially on mycorrhizal fungi. For rhizobia, we did find that the oxygen concentration in the nonfixing soybean nodules fell to half that in the nitrogen-fixing control. High levels of oxygen can damage the nitrogen-fixing enzyme, but normal levels in nodules are low enough to prevent enzyme damage and even lower than optimal for respiration. So maybe the rhizobia weren't getting enough oxygen to reproduce. We can measure nodule-interior oxygen because nodules contain a plant hemoglobin. Like the hemoglobin in our blood, it's involved in oxygen transport and changes color when it binds oxygen. We estimate nodule-interior oxygen from this color change, using a device that Canadian plant physiologist David Layzell and I invented many years ago.[301] It's a bit similar to the instrument used in hospitals to monitor blood oxygen, although the patent wasn't nearly as profitable.

Despite host sanctions, rhizobia that fix little or no nitrogen are still common in some soils. The strains that are most common in soil often fix some nitrogen, but much less than the best strains.[273] We can apply better rhizobia to the seeds, but less-beneficial indigenous strains usually end up outcompeting them and occupying most of the nodules. This means that symbiotic nitrogen fixation is not achieving its full potential as an alternative to nitrogen fertilizer.

Why is that, and what can we do about it? Rhizobial cheaters might persist because, even though they are subject to sanctions when they are in nodules by themselves, they often escape sanctions when they share a nodule with more-mutualistic strains.[302] If so, can we breed legume crops that reduce the number of mixed-strain nodules?

There are also reasons to suspect that sanctions have evolved to be milder than would be optimal for agriculture. Suppose half of a plant's nodules contain a wonderful rhizobium strain, which fixes lots of nitrogen at a low photosynthate cost to the plant. The other half have a mediocre strain that has a higher photosynthate cost, yet fixes less nitrogen. Even so, the fitness benefits of the mediocre strain to the individual plant may sometimes exceed its cost. A plant may have few other options to get the nitrogen it needs for seed proteins. Late in the growing season, roots have already taken most of the available soil nitrogen. Making a new batch of nodules would take too long, and many might get the same mediocre strain. Allocating more resources to the best-performing nodules probably makes sense, and plants apparently do this,[298, 303, 304] but that may not bring in enough nitrogen fast enough.

So plants may continue to support mediocre rhizobia, only imposing sanctions on the worst-performing nodules. This hypothesis is consistent with some additional experiments with soybeans. Kiers exposed nodules to low (rather than zero) nitrogen-gas concentrations, limiting nitrogen fixation to 50 percent of normal. Rhizobia fixing at 50 percent of their potential still reproduced 70 percent as much as rhizobia fixing at full capacity.[305] This is the sort of weak sanction we would expect from individual selection among legume plants.

Even if individual plants increase their fitness by tolerating mediocre rhizobia, why hasn't kin selection in legumes led to stronger sanctions?

Plants that killed rhizobia in nodules with mediocre performance, rather than continuing to support them (and eventually letting them escape into the soil), might produce fewer seeds, but future generations of legumes would benefit. Those future beneficiaries would include the plant's own seedlings. Unfortunately, future beneficiaries would also include its seedlings' competitors. If the beneficiary seedlings are essentially the whole local population, then (applying Grafen's version of Hamilton's rule), we see that $B = P$, so Hamilton's $r$ is zero.

The less widely a plant disperses its seeds, the more it will be surrounded by its own seedlings, moving $B$ closer to $A$ in figure 9.1. But limited dispersal also moves $P$ closer to $B$. On balance, limited dispersal usually doesn't help much.[289] Therefore, kin selection won't necessarily cause plants to impose stricter sanctions than are in their own individual interest.

## Breeding Legume Crops for Stricter Sanctions

What about deliberate human selection for stronger sanctions? Could we breed legume crops that kill almost all the rhizobia in nodules that fix too little nitrogen, rather than just reducing their numbers?

Farmers have much more control over what plants grow in their fields than they do over the bacteria in the soil. We can add great rhizobial strains to soil, but they often die out. On the other hand, only rhizobial strains that can already thrive in a given soil (tolerating any local stresses like drought or salinity) make it into nodules. A soybean or an alfalfa variety with stricter sanctions would select from among those locally adapted strains, enriching the soil with the most beneficial strains over years. If plant breeding can produce crop plants that cooperate with their neighbors, as discussed in the previous chapter, why not make plants that, in effect, cooperate with future generations of plants?

If we understood the molecular details of how plants monitor nodule performance and how they impose sanctions on poorly performing nodules, it might be possible to genetically engineer crops with stricter sanctions. Until then, can we use conventional plant breeding, selecting from among the existing genetic diversity in our crops?

We know that legume genotypes differ in their interactions with rhizobia. For example, Kiers found that three older soybean varieties yielded almost as much when faced with a mixture of good and bad rhizobia as they did with good rhizobia alone. This was not true of three more-modern varieties. The difference was enough to eliminate most of the yield advantage of the modern varieties.[306] We aren't sure that this difference was due to sanctions—plants may sometimes pick the best strains out of the soil, although cheaters may evolve to mimic the recognition signals of better strains—but that's our working hypothesis.

Here's how I would select for legume crops with stricter sanctions.[29,307] As in any kind of selection, we would start with a genetically diverse population of the crop species. We would grow these first-cycle plants in soil containing a mixture of good, mediocre, and bad rhizobia. The easiest way to run these tests would be to use plants in pots. We would record yield for each plant and save its seed. We would then discard seed from any plants with severe problems, but keep in mind that a genotype with low single-plant yield may perform very well as a plant community, as discussed in the previous chapter.

The main criterion for selection, however, would be the performance of a second, genetically uniform *test crop*. This test crop would be grown under low-soil-nitrogen conditions, to make plants strongly dependent on rhizobia for nitrogen. We would record the yield of each second-cycle test plant. We would discard second-cycle seeds, which are all genetically identical, and keep seed only from those first-cycle plants whose effects on soil rhizobial populations benefited second-cycle test plants most. We would repeat this process a few times, until we get a plant genotype that has a strong positive effect on the yield of subsequent crops.

I've assumed that this positive effect on later crops would be due to first-cycle plants with the strongest sanctions enriching the soil with the best rhizobia. But there are other possibilities, such as effects on mycorrhizal fungi. Like rhizobia, mycorrhizal fungi differ in their net benefits to their host plants. Nancy Johnson, an expert on mycorrhizae, has shown that mycorrhizal fungal species that are less beneficial to corn but more beneficial to soybean accumulate in soil under corn, whereas species less beneficial to soybeans but more beneficial to corn accumulate in soil under soybeans.[308] She suggested that this shift in mycorrhizal species may be one reason why corn and soybean have higher yields in rotation (alternating crop species over years) than when the same crop species is grown continuously (year after year). Jim Bever, another expert on mycorrhizae, saw something similar with wild plants.[309] It would be interesting to try my suggested approach with soybean as the first-cycle diverse population under selection, but using corn followed by soybean as test plants. We could thereby evaluate effects of different soybean genotypes on both mycorrhizal fungi (to benefit the corn crop next year) and rhizobia (to benefit the soybean crop in two years).

Other soil microbes also interact with plants. Some are pathogens, causing disease. Others live on the roots, causing little or no harm, and produce antibiotics that suppress pathogens. To the extent that different plant genotypes enrich the soil with pathogens or beneficials, the approach outlined above could generate additional benefits. However, the ability of plants to selectively help the most-beneficial microbes is probably much greater for rhizobia, which are at the plant's mercy once inside a nodule, and for mycorrhizal fungi, which depend on live plants for energy, than for other soil microbes.[273,310]

## Other Opportunities for Improved Two-species Cooperation

Two-species cooperation may also have untapped potential for protecting crops from agricultural pests. Could fungal endophytes protect wheat from fungal pathogens, as they protect cacao leaves? Could ants kill weeds around our crops, as they kill plants that compete with *Duroia hirsuta*? Could ants someday protect orchards or vineyards from insect pests, as they protect acacia trees? Such approaches could offer major benefits someday, but there are also some risks.

Ongoing evolution of agricultural pests and pathogens is a recurrent problem, as discussed in chapter 10. Because fungi have shorter generation times than crop plants do, they can often evolve faster. So fast-evolving endophytic fungi that kill fungal pathogens could be useful in keeping up with pathogen evolution. Maintaining the ability to control pathogens could benefit endophytic fungi that depend on a healthy plant host.

But the potential for mutual benefit is not sufficient to ensure mutually beneficial evolution. Rhizobia would evolve into cheaters, if it weren't for host sanctions. Like legume plants hosting multiple strains of rhizobia, a single cacao leaf typically hosts many different endophytic fungi, creating another potential tragedy of the commons. If one endophyte makes an expensive plant-defending toxin, sacrificing its own fitness for the benefit of the host plant, that will benefit less-altruistic endophytes with which it shares a leaf. So with each successive generation, less-altruistic endophytes might become more common.

At least, that is what we would predict, if benefits from the defensive toxin were equally available to all endophytes in the leaf. Because leaf-defending endophytes haven't disappeared, however, there must be some individual benefit to toxin production, not just a collective benefit. Maybe endophytes are competing with each other within the leaf and making toxins to kill each other, but those toxins also happen to protect the leaf against pathogens.

This sort of individual benefit seems to explain why some bacteria protect plant roots from fungal pathogens, as we suggested in 2003.[310] Subsequent research has shown that, consistent with our predictions, plant-defending toxins made by bacteria are also important in competition among bacteria and in protecting bacteria against predation by protozoa.[311]

If the evolution of plant-defensive toxins is just a side-effect of competition among fungal endophytes, then can we be sure that future evolution of these endophytes will continue to benefit the plant?[273] Is there some way that we can link the evolution of the endophytic fungi more to plant benefit? My guess is that endophytes do have great potential to protect crops, but only if we pay attention to evolution and to possible conflicts of interest between endophytes and their crop hosts.

Similar but more complex issues could arise if we tried to use ants to control weeds. Assume, for the sake of argument, that we can domesticate plant-killing ants much as we have domesticated honeybees, so that they are not aggressive against people. Assume that we have developed a crop that provides the ants with some resource they value and that the ants have been selected to kill any other plants nearby. So far, so good.

But might corn-protecting ants evolve to kill soybeans? How could we select for ants that kill most common weeds but not crops? Dogs can be trained to protect sheep and chickens from wolves and foxes, but can we breed ants that sophisticated? Ant bodyguards might work best for situations where a single crop is grown over a large area for many years, as in an orchard or a vineyard. But the acacia-protecting ants that castrate or prune their own trees should remind us to pay attention to possible evolutionary changes in guardian ants that could cause them to run amok.

Ants are not the only insects known to protect plants. Coffee trees, wild grapes, and wild cotton all provide simple shelters (pits and/or clusters of leaf hairs) for beneficial insects on their leaves. These shelters, known as *domatia*, are typically occupied by (and attract) mites that eat other, harmful mites that attack the plants. They also consume harmful fungi. These domatia are less elaborate than the hollow thorns that house ants protecting acacia trees, but they could presumably be improved by plant breeding.

So far, however, the trend seems to have been in the opposite direction, as most cultivated grape varieties have smaller domatia than wild species do.[312] Similarly, this trait seems to have been lost in domestication of cotton. This may be a case of plant breeders inadvertently discarding some of nature's wisdom.

Further development of domatia as a means of pest control for crops seems promising, but major benefits are not certain. Domatia are sometimes used by plant pests,[313] and the evidence that beneficial mites housed in domatia actually provide significant protection against *herbivores*[314] and pathogenic fungi[315] is still limited. Additional research on the costs and benefits of domatia to wild plants is needed. If benefits consistently exceed costs, then that might be true in agriculture as well. In that case, it might be worthwhile for plant breeders to select for larger domatia in crops (like grapes or cotton) where they may have atrophied since domestication.

My guess is that costs and benefits of domatia will depend on conditions. If so, then we may need to study wild species in different natural ecosystems before we find situations where pest levels and other conditions are similar enough to agricultural systems that we can extrapolate to agriculture.

## Perspective

This conclusion foreshadows a point I will reiterate in chapter 11. Natural selection has improved individual wild plants and animals, not the overall organization of the natural ecosystems where they live. But to understand the adaptations of wild species, including their costs and benefits, we may need to study adaptations in the context where those adaptations evolved: natural ecosystems. For example, many plant adaptations that reduce pest damage can only be understood in the context of multispecies communities, as discussed in chapter 11.

Similarly, the various ways in which wild plants limit losses to rhizobial or mycorrhizal cheaters evolved in natural ecosystems. It is much easier to conduct complex experiments on symbiotic interactions under laboratory conditions, but are we missing some important adaptations that occur only in the field?

Anything we do to manage pests or beneficial species, whether inspired by nature or not, may be countered by their ongoing evolution. For agricultural pests, this point has already received much more attention by other authors than other topics discussed in this book. But I still thought that evolving pests deserved a chapter of their own, so they are the focus of the next chapter.

# 10

## Stop Evolution Now!

UP TO NOW, I have emphasized agricultural insights from understanding past evolution, but evolution continues today. Weeds and insect pests, in particular, can evolve quickly, with major effects on agriculture. For example, over 180 weed species have evolved resistance to a variety of weed-killing herbicides.[316] This ongoing evolution will be the main focus of this chapter.

### Wily Watergrass

We have already developed some effective strategies for slowing the evolution of resistance to our pest-control measures. Consistent with our ideas-from-nature theme, we might get additional ideas from studying wild plants and their interactions with the insects that plague them. But let's start with watergrass, a weed that has evolved resistance to three very different control methods.

Watergrass is a weed that evolved in flooded Asian rice fields, a human-made environment that has been in existence for only a few thousand years. The immediate ancestor of watergrass was barnyardgrass (*Echinochloa crus-galli*), a weed found in nonflooded fields. Barnyardgrass roots are killed by flooding, but its descendant watergrass has air channels in its stems that supply oxygen to its roots. This is a more impressive innovation than anything biotechnology has done so far, but it's only the beginning of watergrass's evolutionary story.

Flooding tolerance was critical to survival in rice fields, but flooding was not the only risk there. One problem for barnyardgrass, even once it evolved flooding tolerance, was that it looks rather different from rice. Both are grasses, so the average city-dweller might have difficulty telling them apart, but to a Chinese rice farmer with a hoe the difference was obvious.

As with most traits, there was some random genetic variation in the appearance of barnyardgrass plants. The plants that escaped the hoe tended to be those that looked more like rice. At first, those survivors probably looked more like barnyardgrass than like rice. However, even a slight resemblance to rice was sometimes good enough, if a farmer had poor eyesight or was working in twilight. A few random mutants among the descendants of these first survivors looked even more like rice, fooling a larger fraction of farmers, and so on.

Eventually, some of these weeds hitched a ride to California, maybe on the shoes of travelers. Have you ever been asked, crossing an international border, whether you visited a farm on your travels? That's why.

In the 1970s, biologist Spencer Barrett and colleagues did a quantitative comparison of visible traits of rice, barnyardgrass, and watergrass in California rice fields. Remarkably, watergrass resembled rice more than it resembled its own recent ancestor, barnyardgrass.[317] This was attributed to selection previously imposed, inadvertently, by farmers with hoes in Asian rice fields.

Today, California rice farmers rarely go out and hoe individual plants. They rely mainly on flooding, tractor cultivation, and herbicides. So, how much has the appearance of watergrass plants evolved in the decades since Barrett's group made their measurements? Has natural selection changed the appearance back toward barnyardgrass? Or has evolution of visible traits wandered off in some new direction? I think it would be interesting to check.

Evolution continues. Although evolution has had millions of years to improve photosynthesis and water-use efficiency, as discussed in earlier chapters, evolution need not take millions of years. For example, weeds can evolve herbicide resistance in only a few years, as shown in a photo I took at a UC Davis "Weed Day" event.

The photo shows watergrass from various California rice farms. Each vertical column of pots in figure 10.1 was treated with a different herbicide, except for the middle column, which was not sprayed. Notice that the watergrass plants grew poorly or died in the two front rows, where sprayed with the herbicides Whip or Abolish, whereas plants in the back two rows of those same columns survived those same herbicides.

The difference is that soil in the two front rows came from two rice fields being farmed organically. No herbicides are used on those farms, so the watergrass seeds in those soils came from plants with no recent evolutionary history of herbicide exposure. Pots in the back rows have soil from farms where herbicides are used. In those fields, only the few mutants that were resistant to those herbicides survived and reproduced. So weed populations evolved resistance on those farms, sometimes in only a few years.

The evolution of herbicide resistance is a widely recognized problem. But it's important to remember that weeds can also evolve resistance to other, nonchemical methods of weed control. For example, watergrass also evolved resistance to flooding and to detection by farmers with

Figure 10.1. Evolution of herbicide resistance in weeds. Research by Albert Fischer and James Hill, of UC Davis, found that watergrass from populations not previously exposed to an herbicide were more susceptible to it. Photo by R. F. Denison.

hoes. Similarly, yellow foxtail growing in alfalfa fields, which are mowed several times a year, evolved to escape mowing damage. It did this through genetic changes that made it grow along the ground, below the cutting height.[318]

## Slowing Evolution

We probably can't stop evolution of resistance to our pest-control measures, but slowing it down can give us time to develop new control methods. To do this, we need to understand the factors that control the rate of evolution.

The rate of evolution depends on how much genetic variation there is within a population. It also depends on the *intensity of selection*. The intensity of selection for a given allele is the extent to which individuals with that allele have greater survival and reproduction than those with other alleles. More intense selection leads to faster evolution.

On the organic farms that were the source of the herbicide-susceptible watergrass in figure 10.1, the intensity of selection for herbicide resistance was zero. It's not that weeds had 100 percent survival there, but that, without herbicides, resistant weeds were no more likely to survive than nonresistant ones. (An herbicide-resistance allele might even impose some fitness cost, in the absence of the herbicide.) The intensity of selection for herbicide resistance was higher on the conventional farms, where herbicides were used. There, herbicide-resistant weeds were much more likely to survive than susceptible ones were. This concept is widely understood.

What is less-well understood is that *increasing* herbicide use may sometimes reduce the intensity of selection. This can be true, for example, for resistance genes that protect against low but not high doses of herbicide. With a high enough dose, such genes confer no increase in survival, so they don't become more common over generations. Higher herbicide doses may endanger humans or wildlife, so it's not necessarily a good option, even if it would slow the evolution of herbicide resistance. But remember this example, when I discuss evolution of insect resistance to the chemical defenses of transgenic crops.

Another reason that increased herbicide use could sometimes slow the evolution of herbicide resistance is that the rate of evolution depends on the amount of *inherited variation* within a population. (One plant may be more resistant than another because it's in a wetter or drier spot, but it's hereditary differences that matter to evolution.) Larger populations usually include more inherited variation in every trait than small ones do. A one-in-a-million mutation is much more likely to occur in a population of a billion plants than in a population of a hundred plants. So, if a farmer can keep weed populations in a field low enough, perhaps using a combination of herbicides and mechanical cultivation, herbicide resistance is less likely to evolve.

The evolutionary importance of population size also has implications for testing new pesticides or pest-resistant crops. Small-scale tests will have lower total pest populations, so fewer potentially resistant mutants, than large-scale commercial production will. So even after a pesticide or transgenic pest-resistant crop has passed initial tests and been approved for sale, ongoing monitoring to detect resistant pests is needed.

With this background, how can we explain the herbicide resistance shown in figure 10.1? At some point, an herbicide-resistant mutant arose (probably, but not certainly, in a field with a relatively high weed population) and survived long enough to make seed. Would a higher herbicide dose have killed this mutant? Possibly, although adding a different herbicide would be even more likely to work. (Similarly, a life-threatening bacterial infection in humans can be brought under control by a combination of antibiotics that work in different ways.)

A nonchemical method, such as flooding (or cultivation, if feasible) would be even better. One reason is that a single mutation sometimes gives resistance to two or more different herbicides, a phenomenon known as *cross-resistance*. Sometimes we can figure out why, after the fact, but it isn't always possible to predict cross-resistance in advance.

Decisions about herbicide use depend on economic and ecological considerations as well as evolutionary ones. A higher herbicide dose may sometimes slow the evolution of resistance, but it will cost more, and it will often have more negative effects on the environment. Using less herbicide isn't the only way to reduce those environmental effects, however. For example, rice farmers in California decreased herbicide pollution of rivers by 98 percent, not by reducing use, but by slowing the movement of water from their fields into rivers, giving herbicides more time to break down.[319]

## Slowing Resistance with the High Dose/Refuge Strategy

To summarize, pests tend to evolve resistance to all of our control measures, not just to chemical pesticides, and using less pesticide is only one of the possible ways to slow evolution of pesticide resistance. Here's a detailed example of a more-sophisticated approach.

The transgenic pest-resistant crops that have been most widely used and studied up to now are those containing protein toxins from the bacterium *Bacillus thuringiensis*, or Bt. This protein kills insects that eat crops containing it. Because it mainly affects caterpillars and has to be eaten to have much effect, it is mostly harmless to other species.

Mostly, but maybe not entirely. One possible ecological risk from Bt crops comes when the pollen they produce lands on other plants, such as the milkweeds that are eaten by Monarch butterfly larvae. Negative effects of pollen from Bt plants on Monarchs and swallowtail butterflies have been reported,[320,321] although these results have been disputed,[322] and any negative effects may be less serious than the alternative of sprayed insecticides.

To the extent that Bt crops have economic and possibly ecological benefits, those benefits would be undermined by the evolution of Bt-resistant pests. Evolution of Bt resistance could also affect those organic farmers who haven't grown Bt crops, but who may sometimes use low-toxicity Bt sprays.

So how can we slow the evolution of Bt resistance? Evolutionary biologists have worked on this problem. They have developed a resistance-management strategy that has been widely adopted and appears to be working reasonably well: the *high dose/refuge strategy*.[323]

The high dose/refuge strategy is based on certain assumptions. The most important of these assumptions is that an insect pest needs two copies of a resistance allele to be resistant to Bt. Like humans, insects have two versions of each gene, one from each parent. Insects with two copies of a resistance allele are designated RR and are assumed to be resistant. Most insects whose ancestors have not been exposed to Bt will probably have two copies of the susceptible version of that gene, making them SS. These are readily killed if they eat a Bt crop. Newly arisen mutants are assumed to be RS, because it's so unlikely that both copies would mutate at the same time. RS insects are assumed to be resistant to low doses of Bt, but killed by high doses.

So the key assumption is that RS insects can be killed by Bt, just like SS insects. So far, this appears to be true, but only if Bt concentrations in the transgenic crop are high enough. Hence the *high dose* part of the strategy. As discussed earlier for herbicides, a higher dose can actually result in lower-intensity selection for resistance, if there is no inherited variation for resistance to high doses.

But even if that assumption is true—that is, no RS mutant has genetic resistance to a high dose of Bt—relying on high doses alone would be risky. What if Bt kills only 99 percent of the RS mutants? Maybe 1 percent of them survived by eating weeds, or something. These RS survivors can't eat the Bt crop without dying, so they aren't really a problem . . . until two of them mate. Then they produce a mixture of 50 percent RS, 25 percent SS, and 25 percent RR offspring. Each RR offspring and its many RR descendants would be resistant to even high doses of Bt, making the Bt crop useless.

So we need to keep any RS survivors (presumably rare) from mating with each other. This is where the *refuge* part of the strategy becomes important. The purpose of the refuge is to supply enough SS insects that the few RS survivors each end up mating with an SS rather than another RS. This is an example of gene flow via flying insects.

For refuges to succeed, they must meet three requirements. First, there must be no selection for resistance there, so that almost all insects in the refuge will be SS. Therefore, the refuge can't have Bt plants or be exposed to Bt spray. Second, survival and reproduction of SS insects in the refuge must be high enough that they outnumber other potential mates in the Bt area. Therefore, the refuge must have plants that can support reproduction of the insect pest, and it can't be sprayed with insecticides that would drastically reduce their numbers. Third, the refuge must be close enough to the Bt areas that SS insects from the refuge will swamp the mating market in the Bt area.

Ideally, refuges will still produce some economic or ecological benefit, despite supporting reproduction of Bt-susceptible pest insects. For example, the refuge could consist of wild plants that the insects can eat, or a crop they can eat without affecting yield too much. Differences between refuge and nonrefuge crops in the timing of planting may sometimes reduce yield losses in the refuge.[324] Still, refuges will often represent a short-term economic loss to farmers.

In recent years, Bt crops have been planted on millions of acres worldwide, with substantial regional differences in land allocation to Bt-free refuges. Not all refuge areas have been deliberately planned. An organic or a conventional farm that chooses not to grow the Bt version of a crop, for whatever reason, may act as a refuge for Bt crops on other farms nearby. Or a different crop can act as a refuge, if it also supports reproduction of the particular pest species.

When these factors are taken into account, the rate of resistance evolution in the field appears to be consistent with mathematical models developed by evolutionary biologists. For example, *Helicoverpa zea* evolved significant resistance to Bt in Arkansas and Mississippi, where refuge area was about 39 percent of Bt-crop area, but not in North Carolina, where the effective area of refuges was about 82 percent.[325]

Even a 39 percent refuge area could represent a significant economic cost to farmers, however, unless the refuge can be planted to a crop that is profitable, while still producing significant numbers of Bt-susceptible insects. So one might expect farmer opposition to refuge requirements. This isn't always true, however, as shown by the history of resistance management in Australia.

## Australian Experience with the High Dose/Refuge Strategy

Previous experience with pests evolving resistance to sprayed insecticides apparently helped convince Australian cotton farmers that they needed to slow the evolution of Bt resistance in cotton pests when they started growing Bt cotton. A committee of farmers, researchers, and government-agency and industry representatives developed a resistance management plan. Initially, farmers themselves called for a 70 percent refuge requirement, although this percentage was later reduced.

Impressively, this reduction was based on science, not shortsighted political pressure from farmers.[326] The scientific justification for reducing refuge requirements came with the arrival of a new variety of cotton that made two different versions of the Bt toxin, rather than just one. Assuming no cross resistance—a big assumption!—the chances of simultaneous mutations giving resistance to both toxins in the same individual insect are very low. A supply of susceptible insects from refuges is still needed, to make sure that individuals with different resistance mutations mate mainly with susceptibles rather than with each other. Otherwise, their offspring could have some resistance to both Bt versions. But, with two resistance genes, the refuge size could be reduced.

The overall economics of Bt cotton in Australia have varied among farms and regions, but there has been an interesting trend over time. The first year that Bt cotton was used, the higher price of seed for Bt cotton outweighed the benefits of reduced pest damage, for a net economic loss. That was followed by three years of break-even net benefits and then by two years with a strong positive economic return.

This is reminiscent of the yield trends some farmers report after switching to organic methods. Lower yields at first are sometimes followed by improvements, which have been attributed to improvements in soil microbial communities.[327] Similarly, trends of decreasing pest populations may have been a factor for Bt profitability in Australia. In both cases, however, increasing farmer experience may also be important. At LTRAS, we showed that first-year organic yields can be as high as those in established organic fields, if both are managed by the same people.[328]

More than 80 percent of Australian farmers listed "protection of the environment" as their main reason to use Bt cotton. Use of sprayed insecticides decreased by 20 to 80 percent. From a short-term ecological perspective, any reduction in sprayed insecticides is probably good. From a longer-term evolutionary perspective, however, spraying any Bt fields that have unusually high pest populations (with a non-Bt pesticide) may be a good idea, as those large populations may contain resistant mutants. Spraying refuges, on the other hand, would reduce the pool of Bt-susceptible insects, making the refuge less effective.

Current refuge options in Australia are flexible, reflecting scientific understanding of differences between refuge and Bt fields. For each 100 hectares of Bt cotton, a farmer can plant either 10 hectares of unsprayed non-Bt cotton, or 100 hectares of sprayed non-Bt cotton (with fewer surviving insects per hectare, more hectares are needed), or as few as 5 hectares of unsprayed pigeon pea. Pigeon pea produces a large number of Bt-susceptible insects per hectare. As long as farmers have to set some land aside as refuges anyway, they often position refuges to avoid spraying pesticides near houses or streams.[326]

The involvement of farmers in development of this resistance-management strategy was key to its success. Still, farmers don't always like being told what to do, even by other farmers. So there are some individual financial incentives, provided by the company that sells the Bt-cotton seed, to encourage everyone to follow the rules. As in the case of the legume sanctions and symbiotic rhizobia,[298] individual rewards or punishments can help to maintain cooperation, when collective benefits fail.

## Area-wide Management Requires Cooperation among Farmers

Cooperation among Australian cotton farmers was key to the relatively successful management of Bt resistance. Cottony cushion scale, an insect pest of California citrus crops, shows how a more individualistic approach can fail.[329]

Beneficial vedalia beetles, predators that feed mainly on the cottony cushion scale insect, were introduced to California from Australia and were very successful at controlling the pest, for a while. Unfortunately, these beneficial beetles are killed by insecticides used against other pests. Farmers who weren't using insecticides often lost their beneficial beetles anyway, when the beetles visited a neighbor's farm and were killed there.

Those insecticide-using neighbors didn't see an immediate increase in their own cottony cushion scale damage, because scale insects were controlled fairly well by the insecticides they were using for other insect pests. But many farmers who had been relying on biological control by the vedalia beetles found that they weren't working anymore, so they started spraying insecticides themselves. Those insecticides then killed still other beneficial insects, which had been controlling other pests, leading to even more pesticide use.[329] This downward spiral illustrates the point that insect pests are often an area-wide problem. Pest-management activities by individual farmers can have big effects, positive or negative, on their neighbors.

In Australia, biological control of citrus scale insects has been more successful. First, one major citrus grower started using biological control. Some Australian farms are huge, and maybe this one farm was big enough that pesticide use by neighbors didn't kill too many of its beneficial insects. Neighbors copied this successful practice, until 75 percent of citrus growers were using biological control, saving money and reducing overall pesticide use by 75 percent.[330] Biological control of scale insects in Australia apparently involved little formal coordination, in contrast to the Bt-cotton example.

Area-wide management of cabbage-family pests, also in Australia, is something of an intermediate case. Diamondback moths were evolving resistance to available insecticides, and public concern about pesticide use was increasing, both of which contributed to widespread farmer interest in better approaches. As in the Bt-cotton case, cooperation among farmers, government agencies, and industry was important.

Farmers switched to Bt sprays, which kill specific pests of cabbage without killing predatory insects that control other pests. But the overall pest-management strategy also included a 3-month "break," when no crops that the diamondback moth can eat were to be grown. Our 2003 paper on Darwinian agriculture proposed a more extreme version of this approach, where no farmer in a region would grow a crop that a particular pest needs to survive, perhaps for a year or more, driving that pest to local extinction.[28] We used this as an example of how humans could design an ecosystem-level pattern not found in nature but potentially beneficial for our purposes.

At first, 70 percent of Australian farmers were using the recommended 3-month break. But then some of them found that a 1-month break reduced pest levels enough on their individual farms, yet let them produce a crop at times when other farms were on break, leaving less competition for markets. Even 1-month breaks might have worked fairly well, if everyone had their break at the same time, but they didn't. So diamondback moths could usually fly from one farm to another and find food.

It has been argued that one key to the success of area-wide pest management is "how to keep the majority of people acting toward the public (that is, their own) long-term good."[330] If there is clear evidence that each individual farmer really increases his own long-term welfare when he acts in ways that promote the long-term public good, then presenting that evidence to farmers may be enough.

But "if everyone did X, we would all be better off" is not always the same as "if I do X, I will be better off." This was one of the key insights in Hardin's paper on the tragedy of the commons.[287] Individual incentives, like those provided to encourage compliance with the Bt-resistance-management program, may be needed. These incentives could include peer pressure.

Although farmer compliance with currently recommended practices is an important challenge, developing new and better methods of sustainable pest management may be just as critical. Once again, nature may be a useful source of ideas.

## Slowing Pest Evolution—Tricks from Nature

There are two different ways in which nature can serve as a source of ideas for improving pest control in agriculture. First, we may get ideas from comparing natural ecosystems, while recognizing that pest control in any particular natural ecosystem has not been improved by competition among ecosystems. Second, we may want to copy some of the pest-defense strategies of individual wild plants, which *have* been improved, by competitive natural selection.

We know that insects that attack wild plants are often eaten themselves by other insects, by birds, or by bats. Similar food webs occur on farms,[331] although we don't always recognize their existence until an insecticide used to kill one pest kills the natural enemies of another pest, as in the preceding citrus scale example.

Because no natural process has optimized food-web structure to maximize the overall productivity or stability of natural ecosystems (chapter 6), uncritical mimicry of natural food webs on our farms would be foolish. However, comparing food webs and their effects on pest damage in a variety of natural ecosystems might help us identify particular food-web features that minimize losses to pests.

The pest-defense strategies and tactics of individual wild plants, on the other hand, have been repeatedly tested competitively and improved by natural selection. Individuals using various genetically programmed defenses competed, directly or indirectly, against other members of their species. Today's wild plants are descended from the winners of many such past competitions, and they inherited the winning strategies and tactics. (Our crops are also descended from wild plants with winning strategies, but they may have lost some of those strategies after humans took more control over their evolution.) So if we copy the pest-defense strategies of individual wild plants, we know we are copying something that worked better, at least in past environments, than a wide variety of alternatives.

In particular, strategies that were quickly defeated by pest evolution would not have persisted, so the strategies that did persist in wild plants must have been at least somewhat evolution-proof. Here are some examples.

Wild potatoes have two types of glandular hairs, or *trichomes*, on their leaves. One type releases a sticky substance when disturbed by insect pests, gluing them in place and preventing them from doing much damage. This type is sometimes found on cultivated potatoes as well. I remember a seminar at Cornell, years ago, reporting that some sprays used to control fungal pathogens inactivate this glue, making potatoes more susceptible to insect damage.

The second kind of trichome is even more interesting. These trichomes release gases similar to the gaseous alarm signals of aphids, which aphids use to warn relatives nearby that they are under attack by predators. Aphids smelling plants' false alarm signals suddenly decide they have urgent business elsewhere, and leave the potato in peace.[332]

Wild-potato genes for making these alarm signals have been identified. This information might be useful to traditional plant breeders trying to transfer this trait from wild to cultivated potatoes. A breeder might want to keep most of a cultivated potato's traits, but incorporate a few genes from a wild potato. One way to do this is to cross wild and cultivated potatoes, then repeatedly back-cross to the cultivated parent, keeping only the progeny that have the desired gene(s) from the wild plant. Knowing what those genes are may make it easier to identify which progeny to keep.

But would cultivated potatoes that make this alarm signal really suffer less damage from aphids? If so, how quickly would aphid populations evolve to ignore the false signal? The fact that natural selection has maintained this trait in wild potatoes suggests a benefit to individual plants in the wild, but things might be different on a farm covered with cultivated potatoes.

If *all* the plants on which the aphids could feed make the alarm signal, that might impose strong selection on aphids to ignore it and to keep feeding on alarm-producing potatoes. This seems similar to plants making natural insecticides (or Bt), thereby imposing selection on pests for resistance to those insecticides.

But maybe not. Aphid responses to their own genuine alarm signals have been maintained by strong natural selection. Aphids that ignored the warning got eaten by predators. So if we could maintain fairly high populations of aphid-eating predators in fields of potatoes that make false alarm signals, the aphids would be caught between two bad options: ignore the signal and get eaten, or continue to respond to the signal and stay away from our potatoes.

Identifying an alarm-signal gene also raises the possibility of genetic engineering to transfer that gene to other crops, perhaps protecting them from aphids. A group at Rothamsted (the agricultural research station mentioned in chapter 2, for pioneering research on long-term sustainability, also known for earlier research on wild potato trichomes[332]) put the wild-potato alarm gene into *Arabidopsis*.[333] As expected, aphids exposed to air around transgenic alarm-producing plants seemed agitated, moving around much more than those exposed to air around control plants. Moving around isn't quite the same as fleeing in terror, however, so additional research is needed to determine whether transgenic plants that make the alarm signal actually repel aphids in the field.

Biotechnology is not the only way to repel pests from crops they would normally attack. In East Africa, farmers have mixed corn or sorghum, which normally attract stem borers, with molasses grass, which repels borers, apparently by producing gaseous chemicals. Borers might evolve to ignore those chemicals, except that farmers also plant Napier grass nearby, which strongly attracts stem borers. Borers therefore lay most of their eggs on Napier grass, but few survive on that host. (How long will it take borer evolution to eliminate this preference for an unsuitable host?) Using such combinations of repellent crops and attractive but lethal "sinks" is known as the *push-pull strategy*.[334]

A somewhat similar approach has been used, also in Africa, to control the parasitic plant, *Striga*, or "witch weed." When *Striga* seeds in the soil detect a host plant nearby, they germinate, grow toward it, attach, and start drawing nutrients from the hapless host, sometimes killing it. Cleverly, someone discovered a crop that released similar chemicals as *Striga*'s usual host, triggering germination and attack, but without allowing actual attachment. Without connecting to a compatible host, *Striga* seedlings soon die. So growing this "trap crop" tends to deplete the supply of *Striga* seeds in the soil, decreasing *Striga* problems over years.

Returning to our alarm-signaling potatoes, there are other ways in which crops might benefit from making aphid alarm signals, aside from just scaring the aphids away. Some wasps that parasitize aphids "eavesdrop" on alarm signals, and use them to track down aphids. Rothamsted researchers found that one parasitic wasp species spent much more time on the alarm-producing transgenic *Arabidopsis* plants than on control plants, looking for the aphids the wasps smelled, but couldn't find.

## Perspective

Mention evolution and agriculture together, and many people think of the rapid evolution of agricultural pests and the development of resistance-management strategies that slow this process. At least, that has been my experience. But ongoing evolution of pests has been only a minor theme of this book, for several reasons.

First, pest evolution has received considerable attention elsewhere.[316,325,326] Second, even perfect pest control would not be enough to meet increasing demand for farm products, driven by population growth and greater use of farm products as chemical feedstocks. We need to increase yield potential for actual yields to keep up with demand. Last, understanding the contrast between long-term, competitive improvement of individual adaptations and the lack of such long-term improvement in overall ecosystem structure is as important to evolution-resistant pest control as it is to improving water-use efficiency.

Recognizing that natural ecosystems have not been perfected by any natural process, however, does not diminish their value as a source of ideas for improving agriculture, and pest-control in particular. This is particularly true when three species (potato, aphid, and wasp, for example) interact. So the next chapter considers some more-complex interactions and what we can learn from natural ecosystems.

# 11

## Learning from Plants, Ants, and Ecosystems

WE CAN LEARN much from natural communities, if we don't mistakenly assume perfection. Natural selection tends to improve the fitness of each participant in multispecies interactions, regardless of the impact on the community as a whole. Understanding this, can we learn how to optimize such interactions in an agricultural context?

### Signals, Cues, and Manipulation

The previous chapter explained how gaseous chemicals that benefit aphids in one context (warning them of predators) can harm them in others (scaring them away host plants). I will start this chapter with an evolutionary perspective on chemicals as either signals, cues, or manipulation. Then I will revisit the fungus-growing ants introduced in chapter 1, focusing on their natural pest-control methods. Last, I will emphasize the importance of natural ecosystems in providing essential context for understanding the sophisticated adaptations of wild species, before applying them to agriculture.

As discussed in the previous chapter, some wasps use aphid alarm signals to find aphids to parasitize. Other wasps respond to volatile chemical "distress signals" released by plants being attacked by caterpillars.[335] Are these really *signals*—that is, are they really beneficial to both sender and receiver? If so, how did they evolve?

The term *signal* is sometimes used loosely. An extreme example can be found in recent use of the term to refer to any chemical released by one microbe that has nonlethal effects on another microbe, even when the same chemical is lethal at higher concentrations. For example, antibiotics made by some bacteria kill other bacteria at high concentrations. At low concentrations, however, antibiotics may induce bacteria to aggregate into layers known as *biofilms*.[336]

Biofilm formation is often seen as a cooperative activity. But my former PhD student Will Ratliff and I have suggested[337] that biofilms formed in response to antibiotics may be analogous to *selfish herds*,[338] where animals try to push into the center of a crowd to escape predators. Is a cell joining a biofilm in response to an antibiotic really doing so to benefit the other cells already there, or to escape from the antibiotic, whatever the consequences for its new neighbors?

It helps to use different terms for information-bearing molecules that have different effects on the fitness of senders and receivers. After all, those fitness effects will determine how production of these molecules, and responses to them, will change under natural selection. The term *signal* should be reserved for messages (including chemicals) that benefit both sender and receiver. A chemical that merely happens to provide useful information to others, but isn't produced for that reason, is better described as a *cue*.[339]

Which term we use can depend on whose fitness we are considering. Aphids may signal to their relatives, but those signals can be used as cues by parasitic wasps. Natural selection will enhance the ability to detect and respond to useful information, so it is not surprising that wasps can detect aphids some distance away. In this case, producing the volatile chemical has conflicting effects on the fitness of the aphids, helping their kin avoid predators, but also increasing the chance of detection by parasitic wasps.

An individual that benefits by producing a chemical message that harms the recipient, typically by providing misleading information, is engaging in *manipulation*. The bolas spider, which makes moth pheromones to lure male moths close enough to catch, is an example.

Natural selection among recipients will tend to decrease their susceptibility to harmful manipulation, perhaps by reducing their responses to certain chemicals. Selection in the species doing the manipulation, meanwhile, will tend to increase the strength of the signal produced, if doing so increases fitness of the sender. This is another example of an evolutionary arms race. In contrast to manipulation, selection will increase responses to useful cues, responses that have also been called *eavesdropping*,[17] and to mutually beneficial signals. Chemicals released by legume roots that attract symbiotic rhizobia are a textbook example of a signal.[340] By attracting rhizobia and housing them in its root nodules, plants gain access to atmospheric nitrogen. A single rhizobial cell founding a nodule can produces millions or billions of descendants inside, many of which may eventually escape into the soil, so responding to plant signals can greatly increase the fitness of a rhizobial cell. Sender and receiver can both benefit, justifying the use of the term *signal*.

Responding to plant recruitment signals may not always benefit a rhizobial cell, however. What if there are more predatory protozoa around roots, and many more aspiring nodule-founders than there are nodulation opportunities?[341] Do legumes, like the farm-owners in *The Grapes of Wrath*, advertise more openings than actually exist, to attract a surplus of applicants? Do rhizobia manipulate each other, overproducing the quorum-sensing (see glossary) signals they use to estimate their own population density, thereby encouraging competitors to disperse to less-crowded areas?[262] The differences between cues, manipulation, and signals will be important as we consider their possible use in improving agriculture.

## Plant Defense by "Bodyguards"

When they are attacked by caterpillars, some plants release volatile chemicals that attract caterpillar-parasitizing wasps.[335] Recruiting wasp "bodyguards" benefits both the plants, which get protection, and the wasps, which get live hosts in which to lay their eggs. With these mutual benefits, these chemicals qualify as signals.

But how did this interaction evolve? It probably started with wasps eavesdropping. Among parasitic wasps, only those that find hosts will reproduce. Therefore, each successive generation of wasps will tend to be better at detecting cues that lead them to their hosts. Aphid alarm signals are an example of such cues.

Another example comes from wasps that parasitize butterfly eggs. Some male butterflies include a chemical anti-aphrodisiac with their sperm, decreasing the chance that a female will mate again, which would dilute the first male's contribution to the next generation's butterfly gene pool. Some parasitic wasps have evolved the ability to detect this anti-aphrodisiac chemical. The wasp uses this cue to find a pregnant butterfly, then follows her—the tiny wasp may even hitch a ride on the larger butterfly—until she lays her eggs, and then parasitizes them.[342]

Given the propensity of parasitic wasps to use all available cues, we can guess how plant signaling to wasp bodyguards may have evolved. Plants wounded by caterpillars would have released some volatile chemical cues just through leakage, even before any natural selection for producing signals. Wasps with behavioral mutations that made them tend to seek out these chemicals were more likely to find caterpillar hosts and reproduce, spreading the alleles for responding to those chemicals.

Once large numbers of wasps began using wounded-plant volatiles as cues, mutant plants that produced more of these volatile chemicals would get more protection from caterpillars, relative to plants that produced less of the same chemical. So a chemical that started as a cue used by wasps became a signal from plants to bodyguards. Some beetles that eat those plants, however, still use the same chemical as a cue.[343]

Some plants have turned signals into manipulation. Certain orchids "cry wolf," releasing volatile chemicals similar to those produced by wounded plants, even when they are not being attacked. These chemicals attract parasitic wasps. During their futile search for caterpillars to parasitize, they pollinate the orchids.[344]

There are probably many more examples of chemical information exchange among plants and beneficial and harmful insects not yet discovered. Can we use what we learn about signals, cues, and manipulation to reduce pest damage in agriculture?

## Biotechnology and Biological Control of Pests

As noted at the end of the previous chapter, advances in biotechnology have made it possible to change the chemical cues released by crop plants, in ways that might enhance pest control. For example, we may be able to increase the ability of crops to attract beneficial predatory or parasitic insects. Would this approach provide more-lasting protection against pests, relative to the current approach of adding insecticidal toxins to crops?

Biological control of pests by predators or parasites that survive and reproduce on a farm is potentially "evolution-proof." Pests evolve ways to hide from their predators or parasites, but that imposes selection for better search strategies by the predators or parasites, because they depend on their prey or hosts to reproduce. The pests may take the lead in this arms-race for a year or two,[230] but coevolution by the predators will eventually catch up.

This coevolution is often missing, however, when we apply beneficial insects that don't survive long on the farm, or when we use crops that make their own insecticides, just as it is when we spray insecticides. For example, crops genetically engineered to produce the Bt toxin impose selection for Bt-resistant pests, as discussed in the previous chapter. But Bt-resistant pests in a farmer's field don't effectively select for improved Bt genes in the crop plants. This is because most farmers purchase seed every year, grown in distant fields where those particular evolved pests are absent. Similarly, beneficial insects that are purchased and applied to control pests on a farm won't coevolve with those pests unless they manage to survive and reproduce on that farm.

If farmers plant seed produced on their own farms, there will be some ongoing selection for pest resistance in their crops. There may also be selection for undesirable plant traits, however, including less within-crop cooperation, as discussed in chapter 8. Letting natural selection work on your farm might maintain pest resistance (and otherwise improve adaptation to local conditions), but it could also lead to wasteful competitive traits like excess height.

What if instead of developing crops that make defensive toxins, we develop crops that send gaseous signals that attract beneficial predators or parasites of crop pests?[333] It would probably be easiest to develop crops that produce such signals all the time, rather than only when those pests are present. An individual plant sending such signals, surrounded by nonsignaling plants, would presumably suffer less pest damage. This could also be true for a small group of plants. If you have a few broccoli plants in a community garden that send out bodyguard-recruiting signals, maybe you could attract beneficial predators and parasites to your plants from your neighbor's plot.

But what happens when all the plants on a farm produce the same signal, even in the absence of pests? Beneficial predators and parasites would then have no useful information to guide them to their prey or hosts. This simpleminded approach to using natural predators and parasites would therefore undermine, rather than strengthen, biological control. Rather than using biotechnology to make crops cry wolf, therefore, I would focus on making sure crop genetic improvement programs don't inadvertently eliminate the natural cues that already enhance biological control.

More sophisticated ways of combining biotechnology with biological control might work, however. If most plants in a field produced a volatile chemical that repelled pests (alarm signals used by those pests, perhaps), while a small patch of plants produced attractive chemical lures, maybe most of the pests could be drawn to the smaller patch. With this *push-pull* strategy,[334] predators or parasites might be attracted to the smaller patch by the abundance of prey or hosts, reducing their need to search the whole field.

Alternatively, biological control insects raised elsewhere could be released directly into these small attractive patches. Remember, however, that predators or parasites raised elsewhere will not have coevolved with potential prey or hosts on a particular farm. If natural selection continuously improves the ability of pests to escape introduced biological-control insects, while purchased biological-control insects stay the same every year, it's pretty clear who will eventually win that evolutionary arms race.

## Pest Control in Ant Agriculture

A related problem may limit the ability of fungus-growing ants to control harmful fungi in their nests. The best-known of these pest fungi is *Escovopsis*, which isn't eaten by ants but attacks their fungal crop. As mentioned in chapter 1, ants apparently use "pesticides" to control *Es-*

*covopsis*. Cameron Currie and colleagues at the University of Wisconsin showed that these pesticides are produced by bacteria that live on the bodies of the ants themselves.[23]

If the pest fungi are killed by a chemical, but the chemical is made by bacteria, is this chemical control (analogous to pesticides) or biological control? The bacteria apparently don't interact directly with *Escovopsis*, so it seems more like chemical control.

Whatever we call it, there are important evolutionary questions here. If *Escovopsis* evolves resistance to the bacterial toxins, will the bacteria coevolve to produce new toxins that overcome that resistance? In other words, are the bacteria more like coevolving beneficial predatory wasps established on a farm, or are they more like purchased wasps that don't coevolve with their target pests? To answer this question, we first need to ask why do the bacteria make these antifungal chemicals.

Is there another tragedy of the commons problem here? Making and excreting antifungal chemicals must have some metabolic cost to an individual bacterial cell. Therefore, if a random mutation in one bacterial cell knocks out antifungal production, that cell should reproduce somewhat faster than nonmutants around it. Within a few days (many bacterial generations), bacterial *cheaters* that don't make the antifungal should displace those that do. This should happen, at least, unless bacteria producing the antifungal somehow benefit preferentially, relative to those that don't.

If the bacteria abandon the production of antifungal compounds, that could eventually let *Escovopsis* destroy the ants' fungal crop, causing the ants to starve. That might be bad news for the bacteria. But as noted in chapter 3, natural selection depends on current conditions, not future consequences, and on individual benefits, not collective ones. Could making antifungals somehow provide a short-term benefit to individual bacteria, outweighing its metabolic cost?

If the bacteria ate *Escovopsis*, there could be strong selection among bacteria to produce the antifungals needed to kill their fungal "prey." This would be similar to biological control, where a predatory insect population evolves to counter evolving defenses in its prey. But these bacteria appear to eat ant excretions, not *Escovopsis*. The chemicals the bacteria produce therefore seem more like nonevolving chemicals (or nonevolving beneficial insects) made or raised in a lab and applied in the field. If lab-raised predators don't evolve to counter prey evolution in the field, what would impose selection for antifungal production in bacteria living on ants?

Recently, a possible explanation for the evolutionary persistence of this antibiotic production was published by Currie's group, although they didn't interpret their results the way I do.[345] They found that the ants' bodies can host *yeasts* (single-cell fungi) as well as bacteria. These yeasts reduce the growth of the bacteria. If yeasts attack or compete with the bacteria, then there would be an immediate, individual advantage to producing antifungal compounds that suppress yeast. Antifungals that suppress yeast might also suppress *Escovopsis*, as a useful side effect.

In the short run, the yeast reduce growth of bacteria, limiting their production of antifungals that suppress *Escovopsis*. That was the conclusion emphasized by Currie's group. Over a longer period, however, selection imposed by yeast could maintain antifungal production by the bacteria, keeping them from losing the ability to suppress *Escovopsis*.

But if selection for antifungal production is imposed by a fungus other than *Escovopsis* (by a yeast, for example), why doesn't that antifungal also harm the fungal crop? Or does it? Ulrich Mueller (formerly Currie's major professor) and colleagues recently reported that bacterial "secretions kill or strongly suppress ant-cultivated fungi,"[346] not just *Escovopsis*. This is what we would expect, if there is no direct selection for activity specifically against *Escovopsis*. The ants can only hope that coevolution of the bacteria with the yeast will maintain effectiveness against *Escovopsis* as well.

Or do ants use other methods? The answer appears to be "it depends." There are many species of fungus-growing ants, which differ in various ways. A group of researchers collected nine of these ant species from a national park in Panama and compared them.[347] Patches of antifungal-producing bacteria were seen (with a low-power microscope) on some ant species, but not on others. The family tree for these ant species has been worked out, and the researchers were able to map the presence or absence of these bacteria onto that phylogeny, as in the transfer RNA example in chapter 3. The last common ancestor shared by all of these fungus-growing ant species apparently hosted the bacteria, but that trait was lost twice, in the ancestors of five of the nine species.

Even bacteria-less ant species aren't helpless against *Escovopsis*, however. Those ants apparently make antibiotics themselves, in their own special glands. They then spread the antibiotics on their gardens and on each other. These antibiotics kill *Escovopsis*, but they can also harm the fungal crop. Therefore, behavioral adaptations (using these chemicals only when necessary) may be as important as the biochemical ability to produce them. This need for caution in applying toxic chemicals applies to human farmers as well. For example, herbicides applied to kill weeds in row crops sometime drift into orchards, harming valuable trees.

The mystery of how natural selection among bacteria that live on ants maintains production of *Escovopsis*-specific antifungals may have been solved: their antifungals are not so specific after all. This may also be true for antifungals made by the ants themselves, but those are used by the ants in ways that carefully target harmful fungi.

Like an individual bacterium, an individual ant may not benefit from the effort needed to kill *Escovopsis*. But unlike a bacterium, a worker ant can have no direct descendants. Her genes reach the next generation only through the reproductive success of her mother, the queen. And her mother's survival and reproduction depends on keeping the colony free of *Escovopsis*. So kin selection is key here.

An interesting correlation seen in the study just described is that the ant species that rely more on their own antifungals tend to have larger colonies. Furthermore, those colonies are also more genetically diverse, because the queen mates with more than one partner.[347] A colony with more genetic diversity might produce a wider range of antibiotics, reducing the chances that fungal pests would evolve resistance to all of the ants' chemical defenses.

Although the queens of some social insects, including ants, have multiple partners today, ancestral state reconstruction suggests that they are descended from monogamous ancestors.[348] With monogamy, Hamilton's *r* for a worker's sister can be greater than for her own offspring. (The *greater* depends on an unusual genetic system found in some insects, which also makes workers *less* related to their brothers.[142]) Therefore, kin selection can favor workers feeding the queen's offspring rather than their own. If the queen had multiple partners, on the other hand, then Hamilton's *r* decreases for sisters. So worker bees and ants might not have abandoned reproduction, if their mothers hadn't been monogamous.[349,350] Once workers lost the ability to reproduce, however, queens were free to take multiple partners, perhaps benefiting pest control by increasing the genetic diversity of the colony.

If monoculture and toxic chemicals (carefully applied?) have worked for ants for 50 million years,[19] can we conclude that these practices are sustainable? Maybe, if our definition of sustainability in a human context is "humans not going extinct." But how often do individual ant colonies collapse? I don't know the answer to this question. And, even if monoculture has met the minimal test of persistence, would polyculture be even better?

Unlike whole ecosystems, ant colonies compete against other ant colonies for resources like leaves. Have there been colonies that practiced polyculture that consistently lost in these competitions? If so, can we conclude that monoculture is more sustainable than polyculture?

Maybe not. The main reason ants use monoculture seems to be the tendency of different fungal genotypes to attack each other when in physical contact. Maybe an ant colony would reduce the risk of catastrophic crop failure if it grew more than one genotype of fungus, but they can't grow them together. (The phrase "separating the sheep from the goats" has nothing to do with a tendency of these species to fight each other, but you get the idea.)

Even growing different fungal genotypes in separate chambers might not work, because droppings from ants eating one fungal genotype trigger adverse reactions in other genotypes.[351] Humans, however, have more options for mixing crops, because our crop plants don't usually have these sorts of hostile interactions.

Another way in which ants protect their crops deserves attention. They grow them underground, physically isolating them from most potential pests. Human farmers can sometimes benefit from a similar approach. Each year, it seems, more of my brother's farm is covered with inexpensive plastic greenhouses. This is partly to keep crops warm in cold weather, but it also reduces some pest problems. For example, his raspberries grown under plastic don't get splashed with mud containing mold spores. If you're in Oregon when raspberries are in season, stop by a farmers' market and taste the delicious results!

New and surprising results on ant fungus farms are being published every year. For example, it was recently reported that although much of the nitrogen ants need is imported in the protein of leaves they harvest, some comes from nitrogen-fixing bacteria in their own gardens.[352] Do ants have any way to favor bacterial genotypes that fix more nitrogen, analogous to the sanctions imposed by legume plants on their root nodule bacteria?

Given how many new things we are learning about ant agriculture, it may be too early to try to adapt their pest-control strategies for use on our farms. An approach that has worked for ants for millions of years might not work for us, but it's certainly worth studying. How and why does it work? What factors are key to its success? To what extent are the challenges faced by ants similar to or different from the challenges faced by human farmers?

To really understand the strengths and weaknesses of ants' agricultural practices, we need to study them as complex communities (ants, crop fungi, *Escovopsis*, two or more species of beneficial bacteria, yeasts, and maybe other species to be discovered), ideally in the context where they evolved: natural ecosystems. It is only in that context that we can fully understand their complex adaptations and determine their relevance to human agriculture.

## Ecological Context and the Value of Natural Ecosystems

Ecological context is likely to be equally important in understanding natural selection's other innovative adaptations in wild species. Those adaptations are coded in DNA, but we can't understand them just by studying their DNA.

As discussed in previous chapters, our justified confidence in nature's ancient wisdom is based mainly on repeated competitive testing by past natural selection. Natural selection operates mainly to improve genes, secondarily to improve individuals and families (like ant colonies), and only sporadically (or as a side-effect) to improve entire multispecies communities. So it might seem that we could preserve natural selection's innovations just by saving seeds, or even just DNA.

But our goal is not just to *preserve* nature's library, but to read it, understand it, and apply it. Often, the understanding we need can come only from studying living organisms in the environments where they evolved.

We might guess, from the DNA sequence of a gene and comparisons with similar genes, that the gene codes for an enzyme that attaches a methyl group to some molecule. But could we determine, just from studying the DNA, or even from studying the living plant in isolation, that the function of the molecule is to attract wasp bodyguards? Probably not, unless we already knew about those bodyguards and which chemicals attract them.

Similarly, if phosphorus-scavenging proteoid roots hadn't already been discovered through field research on plants growing in low-phosphorus soils, how would we decipher the functions of the genes involved so that we could make intelligent decisions about whether to transfer those genes to other plants? DNA and RNA sequencing are powerful tools[353,354]— their costs have dropped enough that I'm starting to use them in my own research—but much of their value will be lost unless we study adaptations in the context where they evolved.

This context-dependence of natural selection's innovative solutions is a rarely recognized reason why it is important to preserve natural ecosystems. Currently, the most widely accepted reason to preserve natural ecosystems is that they provide various *ecosystem services*, such as purification of air and water.[57]

One problem with the ecosystem services rationale is that their economic value can be undermined by changes in land use. For example, the value of pollination services provided by wild bees in Costa Rican forest fragments (totaling 147 hectares) to a nearby 1065-hectare coffee plantation was once estimated at $60,000 per year,[355] providing a strong economic argument for preserving those forest fragments. But then coffee prices dropped and the farm switched to growing pineapple, which doesn't depend on bee pollination.[356] Did natural forests suddenly become much less valuable?

What about endangered species? Rare snail darter fish, once featured in lawsuits that delayed dam construction, don't do much for the ozone layer. Maybe they eat disease-causing mosquitoes, but less-rare fish eat many more. The rarer a species gets, the fewer ecosystem services it provides, undermining the ecosystem services argument for preservation. If we value nature mainly for the ecosystem services it provides, should we let rare species go extinct, and focus on the few species that are endangered despite being abundant?

Not if we see wild species as a valuable source of ideas, as I suggest. Then rare species can be just as valuable as common ones.

For example, researchers studying fish that live in cloudy water have discovered that their eyes process polarized light in ways that let them see farther through the murk, a technology that may be adapted to help people driving or flying through fog.[357] Some sharks have a special microscopic skin texture that greatly reduces drag; this discovery has already been applied to boat hulls, reducing fuel use.[358] Sharks also had countercurrent heat exchangers millions of years before humans invented this energy-conserving device. As far as I know, the snail darter hasn't yet inspired any inventions useful to humans, but that may be because they haven't been studied in enough detail.

If wild species go extinct, we may lose valuable ideas encoded in their DNA. If we preserve their DNA, but destroy the ecosystems where they evolved, we may never be able to figure out all the clever things their genes did. Not all of the problems solved by those genes will be relevant to human needs in agriculture, medicine, or engineering. But one good idea can be worth more than a huge pile of lumber.

I should mention one reason why mimicking the overall organization of a natural ecosystem might sometimes make sense, even with limited understanding of how that organization affects ecosystem productivity, efficiency, or stability. I have argued that the adaptations of individual wild plants have been repeatedly tested and improved by natural selection, while the overall organization of natural ecosystems has not. However, most of natural selection's testing of individual adaptations *took place in natural ecosystems*.

So if we find an individual adaptation that we don't understand, but that works so well that we decide to copy it blindly (perhaps by crossing a crop species with one of its wild relatives), do we also need to copy some aspects of the ecosystem where it evolved, to provide an environment where those adaptations will be most successful? If we don't really understand why a particular adaptation works so well, we might not be able to predict which features of the natural ecosystem the adaptation needs in order to keep working. However, I would be very interested in hearing from anyone who has a good example of such a case.

## Comparisons among Ecosystems

Even if the organization of natural ecosystems hasn't been consistently improved by natural selection or any other natural process, we may learn much from studying them. We shouldn't expect ecosystem-level properties, like the number of species or how they are arranged in time or space, to be optimal. But neither should we expect natural landscapes to be as badly designed as some human-modified landscapes.

In a 2003 paper, we suggested studying natural ecosystems the way an educator might study the educational systems of other countries.[28] Which countries are having the best results? What, if anything, do they have in common? This is a very different approach from asserting that a particular country has the best schools and then copying its system uncritically. Similarly, once we agree on how we want our agricultural landscapes to function, we can look for natural (and agricultural) ecosystems that best meet our performance criteria and try to figure out why those exemplary ecosystems work so well.

It's important to note that natural ecosystems that perform poorly by agricultural criteria may be just as valuable as sources of information on what *doesn't* work as those that are highly productive, stable, and efficient. We might even want to study natural ecosystems that perform especially badly, such as those with particularly severe boom-and-bust cycles of predators and prey, as examples to avoid.

This informational role is particularly important for remnant ecosystems that now occupy little land area. Why do we care how much carbon dioxide a remnant prairie takes out of the atmosphere, on a per-acre basis? There aren't enough acres of prairie left for that to matter to the global atmosphere. The real value of remnant natural ecosystems is as a source of ideas and inspiration, both practical (as emphasized in this book) and aesthetic.

## Perspective

In chapter 1, I introduced fungus-growing ants to show that certain practices that are widely considered unsustainable (monoculture, toxic pesticides, and eating high on the food chain) have persisted in natural ecosystems for millions of years. But mere persistence is a weak criterion, relative to competitive testing. This is our second core principle, from chapter 4. For example, the persistence of wild-rice monocultures in natural lakes doesn't prove their intrinsic superiority to more-diverse plant communities—lakes don't reproduce based on successful competition against other lakes.

But ant colonies do compete against each other, with the winners spawning a disproportionate share of new colonies. So the farming methods we see in ant colonies today have been subject to the competitive testing of natural selection. Therefore, according to the first core principle from chapter 4, we might not expect many opportunities for simple, tradeoff-free improvements in ant agriculture. What can we conclude about monoculture, the use of toxins to control pests, and using animals (or fungi) to convert plants to human food?

Because different strains of the ants' fungi attack each other, converting fungal monoculture to polyculture might not qualify as simple. Ant colonies with polycultures of mutually compatible fungal strains may never have existed to compete against ant colonies that practice monoculture. So direct assessments of the effects of crop diversity, as intercrops or in rotation (see chapter 7), may be more useful than anything we could learn about crop diversity from ants.

As for toxins, their widespread use by plants to defend themselves against herbivores adds to the conclusion from ants—toxins can work. And, in some form, they can keep working for millions of years. In contrast to the apparent sustainability of natural toxins, human-designed toxins are often quickly overcome by pest evolution. So we may have much to learn from ants and plants about the sustainable use of toxins. Keep in mind, however, that our criteria are broader than those of natural selection. We want pesticides that keep working for decades, but without causing cancer or harming wildlife.

Last, if ants eat high on the food chain, should we? Maybe sometimes. The third core principle from chapter 4 is that a diversity of approaches can be a useful form of bet-hedging. This idea is explored in the final chapter.

# 12

## Diversity, Bet-hedging, and Selection among Ideas

IN THIS FINAL CHAPTER, I begin by summarizing my main conclusions. Then I will explain the sort of evidence that would make me change my mind. I also consider the possible risks of following my advice, should I turn out to be wrong. In agriculture, as in financial investments, the best insurance against guessing wrong is bet-hedging, by maintaining a diversified portfolio.

I have argued that the way plant diversity is arranged in natural ecosystems (resembling intercropping) is not necessarily optimal, but that doesn't undermine the importance of diversity itself. Greater crop diversity, optimally deployed in time and space, is our best insurance against famine.

Maintaining or increasing diversity in agricultural practices is also important. This diversity should include both irrigated and rainfed agriculture, and both conventional and organic methods. Keeping both forage-based animal agriculture and production of food for direct human consumption is another important bet-hedging strategy. Greater diversity of on-farm approaches, along with ideas inspired by the adaptations of wild species, could help stimulate greater diversity in agricultural research. That diversity is the raw material for selection among ideas.

It is important to recognize that to maintain diversity, we will have to sacrifice some of the potential benefits of focusing on the few crops or approaches that seem most likely to succeed. But diversity also reduces the risk of catastrophe, should the apparently more-promising approaches fail.

### Darwinian Agriculture in a Nutshell

There is increasing recognition that we need to pay attention to ongoing evolution, particularly that of agricultural weeds, pests, and pathogens. This ongoing evolution will tend to undermine all of our pest-control measures, not just methods based on toxic chemicals. Resistance-management strategies, developed by collaborative teams and consistently implemented by farmers, can slow pest evolution and prolong the useful life of our pest-control methods. Because this topic has been discussed extensively by others,[316,325,326,359,360] however, it hasn't been the focus of this book.

Instead, my central theme is that nature's wisdom is found mainly in competitively tested individual adaptations, in wild species and sometimes still in cultivated ones, rather than in the overall structure of natural ecosystems, which have been tested only by persistence. People who advocate copying ecosystem-level aspects of nature may therefore be choosing the wrong things to copy.

While some agroecologists overestimate the perfection of natural-ecosystem organization, some biotechnology advocates underestimate the perfection of existing individual adaptations. Many of the traits that tradeoff-blind biotechnologists so confidently propose to change have already been improved so much, by past natural selection, that further improvement will be difficult, especially if we ignore tradeoffs.

Prolonged natural selection is unlikely to have missed simple, tradeoff-free opportunities to improve individual fitness in wild species, including the wild ancestors of crops and livestock. Although more-complex improvements may have been missed, our present ability to design complex improvements and to predict their many consequences is limited. Therefore, most near-term opportunities for genetic improvement of crops or livestock will involve tradeoffs. Simplistic, tradeoff-blind approaches to biotechnology, such as constitutive increases in the expression of existing genes (for example, for drought tolerance), are likely to disappoint.

Fortunately, many tradeoffs that constrained past natural selection need not constrain us. A less-bitter cucumber may be more attractive to rabbits, but we can build fences to protect our crops. In crops that are never eaten raw, we could breed for high levels of defensive toxins that are destroyed by cooking.

Breeding for enhanced cooperation is the most promising route to increased yield potential. Because natural selection depends on individual fitness, it has rarely optimized the interactions of groups of plants, or the interactions of plants with partners of other species. But we could do both of these. We may be able to breed wheat plants that work together to shade weeds more than they shade each other, or soybean varieties that selectively enrich the soil with the most-beneficial indigenous strains of rhizobia.

Natural selection hasn't optimized the landscape-scale features of natural ecosystems, such as the total number of species or their arrangement in space and time. As far as we can tell, no other process has consistently optimized these features either. So although it's probably worthwhile studying natural ecosystems to see how they work, with a view toward improving agricultural ecosystems, we shouldn't mimic natural ecosystems mindlessly.

For example, we may be able to design better ways to deploy plant diversity in time and space than nature has. Variations on crop rotation (changes over years, unlike those seen in nature) may control pests better than intercropping the same mixture of species year after year, even if the latter is more similar to what we see in natural ecosystems.

Natural ecosystems are critical, however, for various reasons. They provide essential context for understanding the individual adaptations of wild species. Those adaptations, repeatedly tested and refined by competitive natural selection, are sophisticated solutions to the various problems faced by these species in the past. To the extent that agriculture today faces similar challenges, such as drought, we may benefit by copying the relevant adaptations. Crosses between crops and their wild relatives may restore useful adaptations lost since domestication. Biotechnology may allow transferring genes from wild species into unrelated crops, which could also be useful.

But *copying an adaptation need not always involve transferring genes.* For example, once we're sure that plants benefit from providing shelters (*domatia*) for beneficial insects,[314,315] how might we use this information to improve pest- and disease-control? Even if we could identify all the genes responsible for making domatia, in those species that have them, it's far from certain that transferring those genes to species that lack domatia would produce useful domatia, without harmful side-effects. But maybe we could manufacture something that would serve the same function (pieces of tubing of the right size, perhaps, coated with something that would make them stick to plants), and apply them to domatia-less crops from aircraft. This would be a high-tech analog of the houses some farmers provide for insect-eating bats.

As a more-complex example of copying adaptations without copying genes, what might we learn from studying how plants deal with varying conditions? We assume that plants are reasonably adapted to average conditions in the environments where they evolved, but how can a plant be adapted to both a wet year and a dry year? Sunflowers have a particularly sophisticated solution to this problem.

## Phenotypic Plasticity and Bet-hedging in Sunflowers

There are two basic approaches to the problem of varying environments: phenotypic plasticity and bet-hedging.[361] If plants can change their phenotype fast enough to track changing conditions (growing roots deeper as surface soil dries, for example), then phenotypic plasticity is the way to go. But if a plant can't change phenotype quickly enough, then it may be better to bet-hedge.

A plant could bet-hedge by adopting a low-risk, low-reward phenotype, such as using resources (carbohydrates, or perhaps soil water) conservatively. Conservative water use sacrifices the potentially higher growth of less-conservative approaches—transpiration and photosynthesis are linked—but reduces the risk of using up all the soil water in the plant's root zone and then dying. But some plants also bet-hedge by setting up the equivalent of a diverse investment portfolio, where the collapse of any one stock will have little effect on total portfolio value.

An individual plant has only one phenotype at a time, but it can produce seeds with varying phenotypes, in ways that don't necessarily depend on the seeds' genotypes. For example, a plant might produce some seeds programmed to germinate next year, but bet-hedge by also producing some dormant seeds that won't germinate for two or more years. That way, even if next year is catastrophic, the plant may still have some descendants.

Some species in the sunflower family combine phenotypic plasticity and bet-hedging in a remarkable way. The bet-hedging part comes from producing two different kinds of seeds. Darwin actually mentioned these two seed types as an example of the limitations of natural selection, writing that "it seems impossible that they can be in any way advantageous to the plant."[26] This time, Darwin was wrong.

Some of the seeds are equipped with plumes, aiding long-distance dispersal by wind, while others fall close to the mother plant. Given that the plant itself lived long enough to reproduce—most plants don't, in nature—it must be in a fairly good spot. So a nondispersing seed probably has a better chance of surviving, on average, than a seed that blows with the wind and lands somewhere random. This is particularly true for plants that grow in cities, in isolated patches of soil surrounded by concrete, so some urban plants have evolved lower-dispersal seeds.[362]

But betting all one's seeds on the single most-promising location is the exact opposite of bet-hedging, and that's not what many plants in the sunflower family do. They hedge their bets by making some dispersing seeds and some nondispersing seeds. Maybe one or the other type of seed will survive.

What about the other approach, phenotypic plasticity? They do that, too. The seeds equipped with plumes are those produced in the center of the floral disk, which consists of many tiny flowers. The nondispersing seeds are those produced around the rim of the disk. This difference results in phenotypic plasticity, a beneficial change in the ratio of dispersing to nondispersing seeds with conditions.

When conditions are poor, plants can make only small floral disks, with only a few seeds. As conditions improve, they make larger floral disks and produce more seeds. But dropping ever more seeds in the same place would result in too many daughter plants competing with each other for water and sunlight. Instead, the plant increases the fraction of seeds that disperse.[363]

This increase follows from basic geometry. The number of nondispersing seeds depends on the *circumference* of the flower, which is directly proportional to the diameter. But the number of dispersing seeds depends on the *area* of the central disk, which is proportional to the square of the diameter. So the ratio of dispersing to nondispersing seeds increases with flower size.

In good years, the plants produce larger floral disks. Tripling disk diameter triples the number of nondispersing seeds, but the number of dispersing seeds increases by a factor of nine. The result is that these plants distribute their seeds more widely in good years, rather than wasting resources on a large number of seedlings that would compete with each other for the maternal plant's spot. Darwin should have had more confidence in his own theory of natural selection.

I follow the "sunflower strategy" in investing for retirement, reducing the ratio of high-potential stocks to low-risk bonds as I age, but I always hedge my bets with a diverse portfolio. Similarly, farmers can benefit from some combination of phenotypic plasticity (choosing which crops to grow and how to manage them, based on weather, markets, or predicted pest populations) and bet-hedging (growing two or more crops that are not susceptible to the same pests).

To reinforce an earlier point, we can copy the basic sunflower strategy (bet-hedging plus phenotypic plasticity) without copying sunflower genes. Could the particular geometric mechanism that sunflowers use to link plasticity to bet-hedging inspire loosely analogous risk-management strategies in agriculture? I'm not sure.

The broader role of bet-hedging is discussed in more detail later. The main idea is to reduce negative consequences of guessing wrong, even when that means forgoing some potential gains.

## What if My Proposed Core Principles Turn Out to Be Wrong?

Bet-hedging, the third core principle proposed in chapter 4, is something of an insurance policy against the failure of the first two principles: the importance of tradeoffs and the value of competitive testing. But, like all hypotheses, the first two principles are subject to possible disproof by data that are inconsistent with their predictions. So we need to consider what kind of data would disprove these hypotheses. We also need to evaluate the risks of being guided by these principles, given the possibility that they might someday need revision.

If past natural selection is unlikely to have missed simple, tradeoff-free improvements, then biotechnologists need to pay more attention to tradeoffs. Some tradeoffs may be acceptable, but we shouldn't just ignore them. Those proposing more-complex changes (like the $CO_2$-recycling chloroplasts discussed in chapter 5) have less need to worry that they are reviving something repeatedly rejected by past natural selection, although tradeoffs are still likely to play a role.

But what if this first key hypothesis is wrong? What if someone shows that simply increasing the expression of some existing gene has only positive effects, with no tradeoffs between individual-plant competitiveness and whole-crop performance, and no tradeoffs between adaptation to some conditions versus others? I don't expect this to happen, but even one such counterexample—proved, not just proposed—would undermine the universality of my hypothesis. Even if such a counterexample is eventually found, however, I suggest that the cost of considering tradeoffs is low, relative to the risks of ignoring them.

If someone has a promising idea, I wouldn't automatically reject it just because "if it were a good idea, natural selection would already have done it." Instead, I would suggest a small investment of time and money in evaluating possible tradeoffs. This might involve a one-month delay in starting a multiyear project, some phone calls, and consulting with someone who understands the relevant tradeoffs. If preliminary analysis suggests that tradeoffs are likely to be important, then it may be worth investing additional time and money in quantifying those tradeoffs, before deciding whether to proceed with a major project, such as the development of a perennial grain crop or a new transgenic crop intended to be drought-tolerant. This could reduce the chances of wasting 10 years or a billion dollars, at relatively little cost.

What if my conclusions about the imperfection of natural ecosystems are wrong? I have argued that natural selection has done little to improve the overall organization of natural ecosystems, because they haven't had to compete against each other, the way individual plants do. This doesn't undermine the value of nature as a source of ideas to improve agriculture; it just suggests greater focus. I argue that we should concentrate on the individual adaptations of wild species, competitively tested in the context of natural ecosystems, looking for innovations potentially useful in agriculture. Applying this information could involve transferring one or more genes from wild to cultivated species, using either traditional plant breeding or biotechnology. Or, for example, we might use information about how wild insects identify their host plants to design better traps for insect pests.

Again, what if I'm wrong? The information I found on wild rice and reindeer versus caribou weren't perfect comparisons. What if someone shows that expert human management of wild rice (without fertilizer, for a fair comparison) actually gives lower yield, less stability over years, or more water pollution than harvesting unmanaged wild stands under similar conditions? What if someone shows that expert human management of reindeer herds (without supplemental feed) results in less meat production than hunters could harvest sustainably from wild caribou herds under similar conditions? (I have specified expert management in each case, because economic or other pressures sometimes lead to suboptimal management; presumably, this would be true for farms inspired by natural ecosystems as well.) In other words, what if there are processes optimizing natural ecosystems that we haven't yet discovered?

The opportunity cost of ignoring my advice is a bit more complex in this case, since I'm already advocating seeking ideas in nature, just arguing about where in nature to look. A quantitative analysis of tradeoffs relevant to perennial grains would have been cheap, relative to the cost of a 20-year breeding program with a possibly elusive goal. But what about natural-ecosystem-inspired intercropping? It might seem easy enough to just try it. Hundreds of people have tested intercropping, actually,[171,172] but nobody has done the proper 6-year comparison of the same species in rotation.

One key opportunity cost of mimicking the overall organization of natural ecosystems is actually analogous to a risk I have noted from reliance on biotechnology. In each case, excessive focus on one proposed solution can lead to neglect of other avenues that might be more successful. In particular, a focus on copying landscape-scale properties of natural ecosystems could divert resources from a more-promising approach: studying the competitively tested strategies of the individual wild plants and animals that live there.

I don't expect my first two hypotheses from chapter 4 (namely, the importance of tradeoffs and of competitive testing) to be overturned. But my third proposed principle, on diversity and bet-hedging, seems like a good idea anyway. (Unfortunately, it's hard to test a strategy whose value depends on rare events, like global crop-disease epidemics.) So I don't think we should put all of our agricultural research eggs in one Darwinian agriculture basket. Ignoring evolution would be as foolish as ignoring chemistry or economics, but we can reduce long-term risks by supporting a variety of approaches to agriculture and to agricultural research. There are actually two reasons for this. One is based on bet-hedging, the second is based on analogies with natural selection itself.

## Bet-hedging in Food Production

Bet-hedging is based on tradeoffs. This is true whether we define performance as evolutionary fitness,[364] or world food production, or maximizing benefits from agricultural research. In each case, the approach we guess will work best has some risk of failure. If success of any given approach is uncertain, and we bet all our resources on only one approach, then there's a chance of unacceptable results, such as extinction or famine.

Bet-hedging strategies sacrifice some expected performance under average conditions in order to reduce possible outcomes under unusual conditions to an acceptable range. We sacrifice the best-imaginable outcome to avoid the risk of the worst-imaginable one.

Insurance is a good example of bet-hedging. Insurance companies demand more for fire insurance than their average expected payout. This average payout is what they pay if a house burns, times the chance of that happening. So on average, paying for fire insurance is a bad investment. But many people are willing to pay a premium, to avoid the risk of an uninsured loss.

There are two types of bet-hedging, although the distinction isn't always clear. Conservative bet-hedging chooses a less-risky option, despite its lower expected average success. Choosing to grow wheat rather than corn, despite wheat's lower average yield, would be an example of conservative bet-hedging, where drought is less likely to devastate wheat than corn.

Diversity bet-hedging uses some mixture of strategies. For example, one might plant half a farm to corn and half to soybean, even if corn is more profitable, on average. Because corn and soybean are affected by different diseases, it's unlikely that disease would destroy both corn and soybean in the same year. I will focus on diversity bet-hedging, discussing diversity in food production first and then diversity in agricultural research.

Humans depend on just three crops (wheat, rice, and corn) for a majority of our protein and food energy, either through direct consumption or to feed animals we depend on for milk, eggs, and meat.[33] Given that the world supply of grain in storage would last less than two months,[7] a worldwide failure of any of these three crops could lead to widespread starvation.

The worldwide distribution of these crops makes it unlikely that all of the world's wheat would experience severe drought in the same year. A worldwide disease outbreak may be more likely than worldwide drought, however. There is significant genetic diversity within each of our major crops, reducing the chance that any one pathogen would destroy a crop worldwide. Still, such heavy global reliance on so few crops, all in the grass family, seems foolhardy.

Next on the list, in terms of total world food production, are two more grasses, barley and sugarcane. So a pest or pathogen that attacks all grasses would endanger our top five food sources. Next come soybean, potato, and milk products. Soybean and potato are not closely related to our grass-family crops, or to each other, so they are less likely to be affected by the same diseases. Much of the protein and calories in milk came originally from forage grasses, but legumes like alfalfa and clover are also important forages. Two root crops, sweet potato (not closely related to potato) and cassava, are very important in some countries, although their global production couldn't come close to replacing any of our major crops. There are also many minor crops, from buckwheat to sunflower, that are not grasses or closely related to any of the preceding crops. There are also many more wild plants that could, with enough effort, be domesticated as food crops.

Most of these crops and potential crops have lower yields, on average, than the big three. In some cases, they may even suffer crop failures more often than our major crops do. But if crop failures for these less-popular crops tend to happen in different years than crop failures for the big three, then greater use of minor crops could decrease variability in food production over years. Sacrificing some average yield in exchange for reduced risk of occasional severe food shortages would qualify as diversity bet-hedging.

Ecological definitions of diversity typically require greater balance among species, not just a greater total number of species. Similarly, just adding a few acres of minor crops, scattered around the world, isn't enough. We need to reduce the oligocultural dominance of the big three crops enough that widespread crop failure in any one of them would be survivable. This will be tricky, because so much of our agricultural infrastructure is based on those crops. Also, people used to eating rice or wheat may not want to diversify their diets.

Even if rebalancing our crop-species portfolio would decrease regional or global famine risks, that collective benefit may have little effect on decisions made by individual farmers about which crops to grow. Greater crop diversity on a given farm can benefit a farmer by reducing year-to-year variability in income, but that individual benefit may be outweighed by individual costs. To grow more different crops, a farmer may need to buy more different kinds of equipment and spend limited time developing and maintaining a wider range of crop-specific expertise. Instead of growing more crops, a farmer may reduce risk in other ways, like buying crop-loss insurance, or simply building up bank accounts in good years and drawing on savings in bad years. This reduces risks for the individual farmer, but not for society. The tragedy of the commons strikes again.

If greater crop diversity would benefit society as a whole more than it benefits individual farmers, should we subsidize the production of bet-hedging "orphan crops" that farmers would otherwise neglect? This would be analogous to subsidizing vaccines for infectious diseases. If individuals compare the cost of an expensive vaccine with their individual benefit from immunity, they may forgo vaccination. But getting vaccinated also protects others whom an unvaccinated person might infect. Government subsidies are one way to increase vaccination rates to levels that are optimal for society as a whole.[197]

Could a similar approach be used to increase crop diversity? One potential problem is that politicians may win more votes if they subsidize the crops that most farmers are already growing, rather than providing incentives for farmers to grow the greater variety of crops that would reduce our overall risk.

Growing a greater variety of crops is only one example of diversity bet-hedging in agriculture. For example, variability in total food production might also be reduced if some farmers rely on rainfall to water their crops, others irrigate from rivers carrying melted snow that fell months ago, and still others irrigate from wells pumping groundwater that fell as rain or snow years ago. In contrast, if everyone relies on groundwater, they may collectively pump it faster than it is replenished.

Farmers who sell directly to consumers, through farmers' markets or community-supported-agriculture subscriptions, often grow a wide variety of crops. This is certainly true of my brother Tom's farm. He grows 5 species of fruit tree, 4 species of berries, and 38 other crop species. He also grows several varieties of most species, including 18 varieties of tomato, 12 of lettuce, 8 each of summer and winter squash, 7 of persimmon, and 4 each of fig, kale, and artichokes. Up to some point, farmers' markets and the local food movement probably increase regional crop diversity. But farmers will grow only what they can sell. There are limits to how much we can increase the production of root crops, salad greens, or fruit, if most potential customers are eating meat-centered diets.

## Bet-hedging, Organic Farming, and Animal Agriculture

Animal agriculture does have a role to play in diversity bet-hedging, however, in ways that can interact interestingly with organic farming. Livestock convert only a fraction of the energy and protein in their feed into human-edible energy and protein in milk, meat, or eggs. For chickens, conversion efficiency is roughly 15 percent for energy and 30 percent for protein.[16] (Efficiencies are higher if we count only the fraction of chicken feed that could have been eaten by humans.) So all else being equal, agriculture that feeds grain to chickens will produce less energy and protein than if the grain were eaten directly by humans.

But consider what happens if disease destroys half of the grain crop. Humans dependent on direct consumption of grain have suddenly lost half of their food supply. With chickens in the mix, however, a hungry population can eat most of the chickens and also the grain that had been intended for the chickens.

This comparison depends on certain assumptions, however. If there was enough grain to feed chickens, some of that grain could instead have been stored for use in bad years. Given the conversion efficiency of chickens, a grain-eating population and a chicken-eating population might have similar grain surpluses only if the chicken-eating population had fewer people. Is that likely?

In *Famine in Peasant Societies*, Ronald Seavoy says that in most societies with subsistence farming and expectations of sharing among relatives, populations soon approach carrying capacity, consuming most of the food produced in an average year. This leaves little slack in the system to meet food needs in bad years.[365] When individuals are expected to share with their relatives, incentives to build up surpluses are relatively weak. Why decrease demand by having fewer children, if the cost of feeding them is shared with relatives? Why forgo current consumption for long-term investments (for example, irrigation equipment) to increase supply, if you will have to share any future surpluses? Anecdotal evidence suggests there is at least some truth to Seavoy's generalization, but I don't know whether this has been true of most peasant societies.

To the extent that his generalization applies, do (or did) peasant societies that feed some grain to animals reach a lower carrying capacity than those who eat all the grain themselves? For example, does competition between humans and animals for grain lead to lower birth rates? If so, then animal agriculture could provide a buffer against occasional famines in peasant societies.

In theory, this bet-hedging buffer benefit from animal agriculture could work globally as well, if enough people switch from grain-fed meat to direct consumption of grain when crops fail. This switch could be motivated by increasing meat prices rather than altruism.

But cattle, sheep, and goats don't necessarily need to eat grain. Organic farmer Jim Bender has argued, in *Future Harvest: Pesticide-Free Farming*, that their ability to digest grass and other forage crops makes them especially valuable in organic farming.[366] Most forage crops are perennials, whose extensive root systems reduce erosion and increase soil organic matter more than annual crops do. They thereby improve the soil's ability to store water and nutrients. Also, weeds that infest annual crops tend to die out in perennial pastures, so rotating from crops into pasture for a few years can be a good organic alternative to herbicides or to mechanical cultivation, which can cause erosion.

Grazing animals can provide additional weed-control options. Imagine a wheat field that is particularly weedy this year. By the time the wheat is ready to harvest, the weeds will have dropped billions of seeds, creating weed problems for years to come. An organic farmer could prevent this future problem by mowing the field before the weed seeds mature. But that would mean losing all of this year's income from that field. Cows or sheep to the rescue! They could graze the field, producing meat, milk, or wool from wheat plants and weeds alike.

Some forages, including alfalfa, clover, and birdsfoot trefoil, have additional advantages, especially for farmers who don't use synthetic fertilizers. These legumes can get most of their nitrogen from the atmosphere, via symbiotic rhizobium bacteria in their root nodules. So forage-based animal agriculture is less likely to be disrupted by shortages or price increases for nitrogen fertilizer. In theory, these crops can be grown as green manures, plowed under to supply nitrogen to subsequent nonlegume crops. We did this in UC Davis's long-term research plots at LTRAS. But in practice, few farmers, even organic ones, find it worth the expense of growing these forage legumes, unless they are raising cattle or sheep or selling hay to someone who does.

Could the bet-hedging benefits of some kinds of animal agriculture outweigh the intrinsic inefficiency of animals in converting grain to eggs, meat, or milk? *Efficiency*, remember, always implies a ratio of some desired product to some costly input—it's not a generic synonym for *goodness*—but it can still be measured in various ways. For example, should water-use efficiency include rain, or just irrigation water? The answer depends on what problem you are trying to solve.

If we calculate animal efficiency as the ratio of human-edible food produced to human-edible food consumed by the animals, then sheep eating only forages have infinite efficiency—meat or milk produced divided by zero. Animals eating mostly forages may produce more human-edible food than they consume, even if they also eat some grain.[16] Of course, this raises the question of whether the land and water used to grow forages could have grown human-edible food instead. Sometimes, the answer is no. On moderately steep slopes, pastures usually have much less erosion than grain crops. Perennial grain crops may someday change this story.

My central point is that we should bet-hedge against the risk of catastrophic food shortages by increasing agricultural diversity, even though doing so will tend to reduce food production in an average year. In particular, the inefficiency of food conversion by livestock reduces average food production, but animal agriculture may help buffer the food supply in various ways.

Like animal agriculture, organic farming's biggest contributions may involve bet-hedging. Even if it takes more land to grow the same amount of food by organic methods—this is not always true—organic crops may not fail in the same years that conventional crops do. Because organic farms occupy less than 1 percent of U.S. farmland, their current potential to buffer the overall food supply (for example, against sudden fertilizer shortages) is negligible. But could organic farming expand enough to cover possible shortfalls from conventional agriculture?

One possible limitation on the expansion of organic farming is the limited supply of manure. With only one ton of manure per acre of irrigated farmland,[189] only a fraction of crop nutrient needs could be met by recycling manure. Furthermore, much of the nitrogen in that manure comes originally from synthetic nitrogen fertilizer, applied to grain that is then shipped elsewhere as animal feed. Because organic farms build up soil organic matter gradually, however, they might suffer less yield loss from skipping manure one year than would a conventional farm skipping fertilizer for a year.

Some fraction of our farmland could run on nutrients recycled as manure, if transport problems are solved, but it doesn't seem possible to eliminate nitrogen fertilizer on a large scale, unless we get much better at using nitrogen-fixing crops. This could include both those crops eaten directly by humans (for example, beans or lentils) and forages like alfalfa, converted to milk or meat by livestock. In addition to minor legume crops and forages, there are many wild legumes that might be domesticated. One likely candidate is *Apios americana*, a wild legume that produces an edible tuber.

Organic farming's contribution to bet-hedging could be much greater than we might expect from its small direct contribution to food production. That's because organic farms can be a valuable source of research ideas. This is the same argument I made earlier for preservation of rare species that make little direct contribution to ecosystem services. It's the same argument I made for preserving remnant prairies that are too small to absorb significant quantities of greenhouse gases. Now, since we're talking about ideas, let's move from bet-hedging in agricultural production to bet-hedging in research.

## Diversity for Bet-hedging in Agricultural Research

Using diversity for bet-hedging in food production would increase our use of crops and practices that are less-productive, on average, than some "mainstream" crops and practices, but unlikely to fail at the same time as those mainstream crops and practices. Similarly, bet-hedging in research would require transferring some resources from the few approaches we think are most likely to succeed to a variety of approaches that complement each other, in the sense that the chances of them *all* failing are small.

Currently, most agricultural research funding is going to *biotechnology*, broadly defined to include *genomics* (sequencing whole genomes, rather than focusing on particular genes of interest), *proteomics* (automated analysis of lots of proteins), and *metabolomics* (automated analysis of many molecules other than DNA and protein). The main things these "omics" approaches have in common are that they cost a lot of money, generate a lot of data, and are prone to overselling in both the public and private sector—vapornomics.

Getting more data doesn't always lead to better understanding. For example, the U.S. National Science Foundation allocated $200 million to try to figure out what every gene in *Arabidopsis* does. Although this goal wasn't achieved,[367] we have learned quite a lot about this species. Unfortunately, many of the questions we have about improving crops can't be answered using *Arabidopsis*. Below ground, it lacks the edible tubers, nitrogen-fixing root nodules, and mycorrhizal symbiosis that are key to improving root crops, legumes, and most other species. More recently, the various "omics" are being applied to crop species, but it's not clear to me that these self-styled "hypothesis-free"[368] approaches have increased our actual understanding as much as if the money had been shared with traditional, hypothesis-focused research.

Even if we assume that biotechnology is the most promising single approach to solving agriculture's problems, a bet-hedging strategy would suggest making some of this research money available to a variety of other approaches that are currently starved for funding. In fact, however, these neglected areas may have greater potential than biotechnology, especially when biotechnology ignores tradeoffs.

## Neglected Agricultural Sciences

Traditional disciplines that are now drastically underfunded include *agronomy*, the science of crop and soil management. Would it be better-funded if we renamed it *agronomics* and emphasized higher-throughput approaches, like those based on yield monitors and other computerized instruments mounted on farm equipment? Crop physiology and ecology, traditional plant breeding, weed ecology and evolution, soil microbiology, and agricultural entomology are also being neglected. All of these disciplines have much more impressive track records than biotechnology, whose contribution to average yields may be as low as 2 percent.[116]

Corn yields have increased sixfold since 1939, through a combination of traditional plant breeding and improved agronomic management. Although there have been various attempts to separate the contributions of genetics and agronomy, I agree with Matthijs Tollenaar, a Canadian corn breeder, that "100 percent of the increase in grain yield is actually due to the interaction between genetics and agronomic practices." This has included more-even spacing, a more-than-doubling of plants per acre, and breeding varieties that excel under those conditions.[235] I expect that further improvements will also depend on interactions between genetics and agronomy, requiring continuing collaboration between these disciplines and others, including those focusing on environmental issues.

Improvements in genetics or agronomy in isolation have less potential, but they can still be significant. Here's a specific example from agronomy. At LTRAS, irrigation and adding nitrogen (composted manure in an organic system or synthetic fertilizer in a conventional one) gave annual grain yields four to six times those with no nitrogen inputs and no irrigation: 7600 or 11,500 kg/ha of corn each year (for organic and conventional systems, respectively) versus 4100 hg/ha of wheat every 2 years in our unfertilized and nonirrigated control.[30] (Growing wheat only in alternate years is a time-tested agronomic practice in dry climates, because rainwater and plant-available nitrogen that accumulate in soil during the *fallow* year greatly reduce the chance of expensive crop failures. There wouldn't be enough water for nonirrigated corn, however, even in alternate years.)

What about sustainability? Irrigated and fertilized corn had a statistically significant *upward* yield trend over the first 9 years, whereas nonirrigated and nonfertilized wheat had a statistically significant *negative* trend.[30] By 2007, after 15 years of consistent management, unfertilized wheat yields had fallen to 2500 kg/ha every 2 years, starting to approach the 1000 kg/ha per year typical of preindustrial agriculture.[11] Meanwhile, irrigated and fertilized corn yielded 13,000 kg/ha, roughly ten times as much per year.

My main point here isn't uncritical endorsement of the higher-input systems, especially relative to various alternatives with nitrogen-fixing legumes, which are also being tested at LTRAS. Increasing yields without increasing nitrogen fertilizer use is a major focus of agronomic research worldwide.[369] But results from LTRAS do show that yield differences due to crop management are huge relative to gains from biotechnology, at least so far.

The importance of agronomy is also apparent in yield differences among countries in Africa (1100 kg/ha for corn in Nigeria, versus 2900 in South Africa and 7700 kg/ha in Egypt), although climate and soil type also contribute, and in *yield gaps*, the differences between yields with expert agronomic management and average yields in the same country.

Crop physiology and ecology are scientific disciplines whose contributions to crop yield and resource-use efficiency mostly depend on the information they provide to agronomists and plant breeders. Chapter 8 is full of examples. My former major professor, Tom Sinclair, has recently reviewed the role and contributions of crop physiology and the limitations of tradeoff-blind biotechnology.[115,370] Chapter 10 gives a few examples showing how understanding the ecology and evolution of weeds and insect pests remains an essential adjunct to biotech approaches. A bet-hedging approach to agricultural research would restore funding to all of these traditional disciplines, even if that reduces funding for biotechnology.

What about new approaches, inspired by nature? These more-traditional disciplines have been more receptive to input from evolutionary biology, in my experience, than biotechnology has. Additional funding for research on the adaptations of wild species would certainly be worthwhile, especially for species that might soon go extinct, taking their potentially informative adaptations with them. This will be particularly true if we can bridge the communications gap between agricultural researchers and those studying wild species.

Many of those now studying wild species express little interest in agriculture, except when it threatens their favorite species, but targeted research grants would probably inspire some of them to think about agricultural applications. For example, it's amazing how many plant scientists are working either on *Arabidopsis* or biofuels. But that's where most of the money is.

Some scientists are already trying to use ideas inspired by nature to improve agriculture. As explained in chapter 7, I have some doubts about perennial grain crops. But do we really want to focus 100 percent of our grain-breeding efforts on annuals? As long as they pay attention to tradeoffs—a recent e-mail from David Van Tassel, working on perennial grains at the Land Institute, suggests that they are—they might come up with something useful.

This may sound like weak praise, but it's really all we can say about any long-term research project, including my own. Every project has opportunity costs, however, including the other projects the same people could be working on instead. The perennial grain project was apparently inspired by wild prairie species. If progress on perennial grains stalls, as I expect, could one or more of the five scientists working on perennial grains at the Land Institute find another project, also inspired by prairie species, to work on?

It should be clear that I think we should move some of the taxpayer money we spend on agricultural research from biotechnology into the various fields just discussed. But this transfer could, in theory, be taken too far. True bet-hedging would require some ongoing support for research on biotechnology.

Biotechnology approaches that recognize tradeoffs (like the potato example in chapter 8) or those involving such radical innovations that they haven't been tested by past natural selection (like the $CO_2$-recycling example in chapter 5) are perfectly consistent with Darwinian agriculture. They are worth considering on their own merits, even apart from bet-hedging.

## Natural Selection among Ideas

Up to now, I have advocated increasing the diversity of agricultural research approaches as a form of bet-hedging. But diversity can also play another important role. Remember that the rate of improvement by natural selection depends on genetic diversity. The rate of evolution also increases with the strength of selection. Can a process similar to natural selection help us identify the best ideas for improving agriculture? To maximize the efficiency of this process, we need a diversity of approaches to select among, as advocated earlier. But we also need strong selection.

What are we selecting among? Ideas. An idea is an example of what Richard Dawkins called a "meme . . . a unit of information residing in a brain."[5] A useful term, perhaps, having some analogies with gene. We can imagine selection among memes in an audience listening to a debate, or *meme flow* from one field of study to another. But the process of copying a meme from one brain to another is much less accurate than the process of copying a gene from parent to offspring.

For example, the *selfish gene* meme in my brain could be translated into words as "the alleles that become more common over generations are those that act *as if* they were selfishly trying to increase their abundance in the population, at the expense of alternative alleles. For example, they may trick the bodies in which they reside into helping others likely to have the same allele." But, in other brains, the selfish gene meme has mutated into "we humans are selfish because of our genes—it's natural!"

A few years ago, I heard a talk by a graduate student who had been studying how farmers adopt new practices. She focused on a program in which various experts worked with farmers to improve the sustainability of orchards. The farmers got paid for attending meetings and received subsidies for practices that the experts considered more sustainable, such as using cover crops.

The program was apparently a big success at first, if we define *success* as farmers following the experts' recommendations. (An alternative definition would be based on the actual consequences of following those recommendations. For example, was there less nitrate pollution per pound of almonds produced?) But when the meetings, payments, and subsidies stopped, many of the farmers went back to their previous ways of doing things.

The student giving the seminar recommended that in the future, enrollment in such programs should be limited to people who were already strongly committed to using cover crops and biological control of pests. I drew exactly the opposite conclusion. In this program, selection among ideas had been too weak, not too strong.

Farmers readily adopt new practices that work well for them. So the only reason they would abandon cover crops, once the meetings and payments ended, is that they weren't working well. Maybe the cover crops suppressed weed growth, but competed with trees for water. Or maybe they provided habitat for harmful insects, as well as beneficial ones.[371] Or maybe managing the cover crops just took too much time, relative to their benefits.

Apparently, the farmers in the pilot project were too polite to point out problems with the approaches recommended by the experts. In other words, the intensity of selection among ideas was too weak during the pilot program, allowing bad ideas to survive. I don't mean to imply that cover crops are necessarily a bad idea. Maybe they just needed some fine-tuning, which the program didn't manage to provide.

How could this program have increased the intensity of selection among ideas? My recommendation would have been to *recruit farmers who were skeptical* about cover crops (or whatever), so that the experts would have been challenged to address their concerns. If the problems were solved, the farmers would have continued using cover crops, and rightly so. But if the experts weren't able to help farmers benefit from using cover crops, then they would have stopped using them, and rightly so. Rather than trying to persuade farmers to use methods that sound sustainable but don't actually work, let's help them develop methods that do work!

Once the pilot program ended, farmers were subject to a rigorous selection process, driven by real-world constraints like profitability. Those who make their living by farming—not everyone who calls himself a farmer meets this criterion—quickly abandon approaches that lose money. Those who persist with unworkable approaches soon end up selling their land to farmers with more profitable approaches.

This selection-like process, Adam Smith's "invisible hand" of market forces, doesn't always produce optimal results, however. There are at least two reasons for this. First, conditions change. Increases in fertilizer prices, for example, could increase the economic value of nitrogen-fixing legume cover crops. Genetic improvements in the cover crops themselves (stronger sanctions against mediocre rhizobia, perhaps?), or in the methods or equipment used to manage them, could make a formerly impractical method useful. Yet farmers may be reluctant to give cover crops a second chance, especially if they were oversold in the past.

Second, farming practices that are most beneficial to individual farmers may not always provide the greatest benefit to society as a whole. As Cornell economist Robert Frank noted, "Competition . . . sometimes guides individual behavior in ways that benefit society as a whole. But not always."[372] Note the analogy with natural selection. Competition among plants has given us innovations like C4 photosynthesis, which benefit both individual plants and their ecosystems, but also innovations like wasteful excessive stem growth.

Cover crops could perhaps represent a net cost to individual farmers, yet a net benefit to society. For example, cover crops can reduce erosion. Some of the benefits of reduced erosion (for example, maintaining soil fertility) go to the individual farmer, but others benefit society as a whole (for example, less pollution of rivers with silt or agricultural chemicals). We may need to provide farmers with incentives, such as taxes or subsidies, that align their individual interests with those of society as a whole.

Taxes and subsidies can be effective ways to achieve societal goals. But subsidies for specific practices that are only *thought* to result in desired outcomes can backfire. For example, European countries spent almost two billion euros to subsidize agricultural practices intended to increase the diversity of birds, insects, and wild plants on farms. But a comparison among 78 pairs of fields in and out of the program found no increase in diversity of either plants or birds. Fields managed to help wading birds, in particular, were avoided by these birds, apparently because lower fertilizer rates reduced food availability.[373]

## Focus on Outcomes

What should we do when measures intended to protect wildlife or promote other societal goals turn out to have the opposite effect? Scientists would probably suggest changing laws and regulations as new information becomes available. But politicians and bureaucrats can be reluctant to admit that measures they championed have been ineffective or harmful.[374] So rather than promoting particular practices, we should *link taxes and subsidies as directly as possible to actual outcomes.* (A similar approach has been suggested to encourage doctors to focus on outcomes for patients rather than maximizing paid procedures. Some people have even suggested that teachers should be paid more if their students learn more, relative to other teachers' students with similar backgrounds.) With subsidies based on outcomes, farmers will be motivated to use whatever methods work best to reduce pollution or promote wildlife. New scientific results will be put to use to meet performance goals linked to subsidies, rather than condemned as "stupid politicians changing their minds."

One possible problem with subsidies or taxes linked to outcomes, rather than practices, is that outcomes can be harder to measure. It may be easier to enforce timing restrictions on hay harvests, which may threaten nesting birds, than it is to count actual bird nests. But sometimes, outcome-based approaches can be simpler and easier.

For example, most economists agree that a tax on the release of carbon dioxide into the atmosphere would be the most-efficient way to reduce the release of carbon dioxide into the atmosphere. (Countries with balanced budgets could then reduce other taxes.) This tax would be included in the price of gasoline or coal and passed on in the price of anything, including electricity, made using these resources. Energy prices would rise, making alternative energy sources (and conservation measures, like installing insulation) more economically viable. A carbon tax would help the most-practical energy alternatives float to the top, thereby reducing greenhouse-gas production at the lowest societal cost. This contrasts with government regulations or subsidies to promote particular practices.

If it's cheaper to heat a house with electricity from solar cells than to add a south-facing window (in northern temperate regions) and insulation, fine. If "high-diversity grassland biomass"[375] can be converted to motor-vehicle fuel more cheaply than other alternatives, such as stalks and leaves from corn already harvested for grain, great! Let ethanol compete with electric cars charged by electricity from windmills. In agriculture, carbon taxes could promote more-efficient tractors or reduced plowing, better-insulated greenhouses or using heat from composting manure, more-efficient irrigation pumps or growing crops that need less irrigation, whatever works best.

And what about claims that raising lambs in New Zealand and shipping them to England actually saves energy, because the lower energy cost of production in New Zealand outweighs the energy cost of shipping?[35] With carbon taxes high enough to significantly reduce energy use, we wouldn't have to argue about it; market forces would pick the most energy-efficient options.

Nitrate pollution offers another agricultural example. Overuse of nitrogen fertilizer can lead to pollution of wells,[376] used for drinking water by farmers and others, and rivers, eventually affecting fisheries in the Gulf of Mexico.[377] Some excess nitrogen can also be released as nitrous oxide, a greenhouse gas that is about 300 times as harmful as carbon dioxide, on a per-molecule basis. So we don't want farmers to use more nitrogen fertilizer than they need for optimum crop yields.

On the other hand, we don't want them to use *too little* nitrogen either, although synthetic fertilizer may not always be the best source. If nitrogen deficiency limits crop yield to half of its potential, then we need to use twice as much land to produce the same amount of food. The total amount of water evaporating from soil and through leaves doesn't decrease much with decreased crop growth, so nitrogen-deficient crops would also use twice as much water to grow the same amount of food, cutting water-use efficiency in half. Similarly, herbicide use may increase with nitrogen deficiency, because a nitrogen-deficient crop doesn't shade weeds as effectively. This was evident at LTRAS, where the weediest wheat plots were those without nitrogen inputs.

If farmers cut back too much on nitrogen, then we would need to consider global implications as well. Less food would be produced, so food prices would rise. This would create an immediate hardship for low-income people, but rising prices also tend to stimulate agricultural production. So a reduction in nitrogen use in one country might make it more profitable to clear forests or drain wetlands for agriculture in another country.

Environmentalists have noted that diverting grain into biofuel could increase world grain prices and thereby stimulate land clearing for agriculture.[378] But we have not always recognized that grain shortages from other causes, such as using too little nitrogen, could have similar effects. Such indirect effects on land conversion can be complex and hard to predict,[61,63,379–381] but they strengthen my conclusion that using too little nitrogen fertilizer could sometimes be as harmful to the environment, albeit indirectly, as using too much.

So here's the problem. Society as a whole benefits when farmers meet the nitrogen needs of their crops, but without using too much fertilizer. But society as a whole has no clue how much nitrogen is needed on a particular field in a given year.

## Stronger Selection among Nitrogen-management Ideas

A tax on nitrogen fertilizer, superficially analogous to a carbon tax, might seem an obvious solution. But I am suggesting something a bit more sophisticated, based on the overall *nitrogen balance* of a farm. In the short term, some of the nitrogen a farmer adds as fertilizer may accumulate in the soil, rather than being taken up by crops or polluting nearby rivers. Similarly, also in the short term, nitrogen released from soil organic matter may meet crop nitrogen needs for some years, without additional nitrogen inputs. Short-term student projects that get this result may conclude, incorrectly, that nitrogen inputs aren't needed for sustainability!

But nitrogen balances are different over decades. That is why LTRAS was designed as a 100-year experiment. Over the long-term, nitrogen removal in the protein of harvested grain cannot exceed nitrogen inputs from all sources. This is the law of conservation of matter, applied with manifold force to nitrogen. On the other hand, any long-term excess of nitrogen inputs over nitrogen removal in harvested crops will result in nitrogen losses to the environment, most of them harmful. Those losses include nitrate *leaching* down through the soil into the groundwater that supplies wells, surface runoff carrying various forms of nitrogen into rivers, and release of the greenhouse gas nitrous oxide.

Even the best farmers can't completely eliminate such losses. But it is those polluting losses, not fertilizer use itself, that we want to reduce. (We do want to minimize energy used in fertilizer production, but that would be covered already by a carbon tax.) So could we somehow tax nitrogen loss to the environment?

Maybe. We could tax nitrogen fertilizer inputs to the farm, but then provide each farmer with a credit for all the nitrogen that leaves the farm beneficially, in farm products. Because 1000 kilograms of wheat contains 23 kilograms of nitrogen,[16] every 1000 kilograms of wheat sold off-farm would generate a 23-kilogram nitrogen-tax credit. A feedlot buying that wheat would pay a tax on the nitrogen content of the wheat, but receive nitrogen-tax credits for nitrogen sold in meat or milk protein. The feedlot would also get a credit for nitrogen in any manure they sold (or donated) for use on a farm, giving an economic incentive against "accidentally" releasing the manure into the river. Closing the cycle, the farm receiving the manure would pay the nitrogen tax on the manure, minus a credit for the nitrogen in farm products sold.

So everyone would have strong incentives to use as much nitrogen as they need, but no more. Everyone would benefit, economically, from ensuring that the nitrogen ends up in useful products, like grain or milk, rather than polluting water or air. To avoid increasing the overall tax burden on farmers, I would suggest reducing other taxes, like those on land. But it's worth noting that farmers already pay extra taxes on nitrogen fertilizer in some regions. What's different about my approach is that they would be taxed only for nitrogen that ends up causing pollution.

With agricultural taxes and subsidies based on actual results, and advice from agricultural researchers and extension specialists, farmers could probably make better crop-management choices than politicians. We just need to give them economic incentives that align their interests with those of society.

Ongoing research is needed, however, to develop better options than those available today. For example, agronomists have long known that splitting fertilizer applications (some at planting, some later) can reduce pollution and increase yield, relative to applying fertilizer all at once.[11] More recently, this approach has been reinvented by ecologists.[382] Ideally, the amount of nitrogen applied in the later application should be adjusted, based on crop nitrogen needs and soil-nitrogen supply at that time. Both variables vary across a field, however, so different parts of a field should get different amounts of nitrogen. Adjusting the amount of fertilizer delivered to each part of the field is getting easier, with today's computer-controlled application equipment. But how does the equipment know how much fertilizer to apply where?

One promising approach is to get nitrogen-status information from the plants themselves, based on inexpensive measurements of leaf color.[383] This is faster and less expensive than taking lots of soil samples and analyzing them in a lab. But let's not ignore nature's wisdom. Can we breed crops that report their nitrogen needs more accurately, perhaps using genes from wild plants?

## Darwinian Selection among Ideas

Let me summarize my proposed Darwinian approach to finding the best ideas for improving agriculture. We need to generate as diverse a bunch of ideas as possible so that our selection process will have material to work with. This will require rebalancing research funding among molecular, physiological, ecological, and evolutionary approaches. It will require research on organic as well as conventional farms, recognizing that ideas from research on either kind of farm may benefit the other, just as research on irrigated farms may benefit those without irrigation, and vice versa. We also need more research on wild plants, and greater efforts to relate that work to agriculture.

Then we need strong selection, to advance the best ideas. Currently, market forces provide strong selection for short-term profits. This form of selection is useful in eliminating wildly impractical ideas (except those subsidized by credulous investors or government agencies), but it's incomplete. We need selection that advances the long-term interests of society: ensuring an adequate food supply even in bad years (not just on average), keeping pollutants out of our water and air, and (though we may argue over details) preserving wild species and natural ecosystems. Carbon taxes and the nitrogen-budget approach described earlier are presented as examples of approaches that, combined with market forces, could impose such selection.

Natural selection among individual plants doesn't always favor those that play well together. Tall plants that waste resources on stems can crowd out shorter plants that invest more in grain. Similar problems may occur in research. Big labs may overshadow small labs, even when the latter have better ideas. Researchers who were more successful getting grant money in the past will have more students and postdoctoral researchers, generating preliminary data needed for their next grant proposal. Rich labs get richer over time through this process of *cumulative advantage*.[384]

Maybe famous researchers' ideas are better, on average, than those of less-famous ones. But are their third-best ideas better than the less-famous researchers best ideas? Does a lab with ten students and ten postdocs (each getting, at most, 5 percent of the lead scientist's time) produce ten times as many breakthroughs as a lab with one of each?

Apparently not. A comparison by Jeremy Berg, of the U.S. National Institutes of Health, found that the labs with the most funding didn't publish more papers than those with half as much funding. Furthermore, papers from the best-funded labs were actually less influential, as measured by citations per paper. An article describing this result was titled "Middle-sized labs do best,"[385] but the smallest labs actually published four times more papers per dollar, with only slightly lower citation rates per paper. This is the same sort of diminishing returns we see in nature, or in agriculture. Giving a nitrogen-starved crop a little nitrogen will increase its growth and water-use efficiency a lot, but give the crop too much nitrogen and most of it will be wasted.

So concentrating our limited research money on a few big labs may be a bad idea. As success rates for grants have dropped below 10 percent, the National Science Foundation has supposedly become less willing to fund researchers who already have plenty of grant money. This seems like a good idea to me.

Even if a few researchers seem to have most of the best ideas, a bet-hedging strategy might favor spreading the money around more than it is at present, at least in the United States. Canada tends to do this already, so it would be interesting to compare the research contributions of Canada, per dollar spent, with those of the United States.

As natural selection eliminates the alleles that contribute least to fitness, the overall genetic diversity of a population decreases. Similarly, rigorous selection among ideas for improving agriculture will reduce the very diversity of ideas that selection needs to work. So we need a constant supply of new ideas.

Populations acquire new genetic alleles by mutation and via gene flow. Alleles arriving from elsewhere, via gene flow, aren't necessarily adapted to local conditions, but they have presumably survived natural selection's repeated testing back home. Human-mediated gene flow from wild species (for example, by plant breeders crossing crops with their wild relatives), and meme flow from researchers who study those wild species may serve a similar function, replenishing the gene pool and the meme pool for improving agriculture.

## Perspective

I hope that this book has convinced you that the challenges facing agriculture are important enough and difficult enough that we need more smart people working on them. Perhaps you, or someone you know, will be among those people.

You might use some combination of biotechnology and traditional plant breeding to improve the genetics of our existing crops or to develop new crops by domesticating wild species. If so, I hope that my discussion of tradeoffs will be helpful. If you are a farmer, you may already experiment with different crops (potentially diversifying agriculture in your region) or crop rotations. Are you willing to share what you have learned?

Maybe you will study natural ecosystems or wild species for their own sake, but keep your eyes and mind open for information that might be used to improve the efficiency or sustainability of agriculture. If so, I hope you will recognize when the organization of natural ecosystems undermines overall performance, by criteria relevant to agriculture, as well as cases when that organization might enhance productivity or stability.

Perhaps you will have an opportunity to influence research funding by governments or foundations. If so, I hope that this book has convinced you that ecology and evolutionary biology (and their applied cousins, agronomy, plant breeding, and so on) deserve more support, even if that means less money for biotechnology. If you have an opportunity to influence taxes and subsidies that affect farmers, I hope you will promote incentives linked to desired outcomes, rather than dictating particular practices that may turn out to be suboptimal.

You don't need to be a full-time scientist, farmer, journalist, or government official. Carol Deppe has argued convincingly that individual farmers and hobbyists can make significant contributions to plant breeding, especially with crops that are neglected by major seed companies.[386] Bloggers and those who comment on blogs sometimes raise important points that have been missed by professional journalists.

Simply changing your individual lifestyle, however, isn't likely to have much effect. You might be able to reduce your personal contribution to global warming by bicycling to the grocery store, where efficient trucks deliver food by the ton, rather than driving to a distant farmers' market or pick-it-yourself orchard for a few pounds of produce. But then someone else will use the gasoline you saved, perhaps for less-beneficial purposes. If you really want to make a difference, you need to influence significant numbers of other people. And that means, at least occasionally, working in the realm of innovation and ideas. I hope that this book will inspire and equip some of you to take up that challenge.

# Acknowledgments

I THANK Richard Dawkins for calling my attention to Muir's interesting work with chickens. The influence of Dawkins's books, particularly *The Selfish Gene* and *The Extended Phenotype*, should be obvious. University of Nebraska agronomist Ken Cassman provided many useful insights and suggestions. Melissa Ho, then a teaching assistant in my crop ecology class at UC Davis, introduced me to the paper critiqued in chapter 7. I am grateful for specific information and comments on the text from many people: Paul Bloom, Jeremy Burdon, Lee DeHaan, Fyodor Kondrashov, Jouko Kumpula, Igor Libourel, Daniel Marcum, Merry Rendahl, Ed Swain, Maththijs Tollenaar, Jacob Weiner, Xiangming Xu, Myron Zalucki, my wife, Cindy, and my brothers, Tom and Glenn. My sister, Becky, my parents, my mentors Tom Sinclair and Bob Loomis, and various colleagues, notably Tom Kinraide, have influenced me more than they may realize. Chapters 1 through 9 benefited greatly from comments from UC Davis entomology professor Jay Rosenheim and population biology graduate students Bonnie Blaimer, Gideon Bradburd, Nicholas Fabina, Ryan Runquist, Clarissa Sabella, Chris Searcy, and Alisa Sedghifar in a seminar class he led on an early version of this book. My own graduate students Will Ratcliff and Ryoko Oono helped clarify my thinking on some important points. My research on agricultural sustainability and the evolution of legume-rhizobia mutualism has been supported by the U.S. National Science Foundation, the U.S. Department of Agriculture, and the Kearney Foundation of Soil Science. My appreciation to each of these people and organizations does not imply their endorsement of my conclusions.

# Glossary

**adaptation** — A trait that increases fitness in a given environment and that results from past selection.

**agroecologist** — One who tries to apply ecological principles to agriculture, sometimes including the questionable principle that natural ecosystems have somehow been perfected over millennia so that agriculture should copy their overall organization.

**allele** — Alternative version of a gene, such as different alleles for eye color.

**amino acid** — One of the twenty building blocks (each coded for by a different DNA triplet of three nucleotides) strung together into proteins.

**ancestral-state reconstruction** — Determining the most-likely genotype or phenotype of one or more of the ancestors shared by a group of living species.

**aquifer** — Underground layers containing water, typically in sand, gravel, or permeable rock. Water removed via wells comes from aquifers, but water in some aquifers may also flow, usually slowly, into rivers, lakes, and oceans.

*Arabidopsis* — A small weedy plant often used for basic research.

*Bacillus thuringensis* (Bt) toxin — An insect-killing toxin produced by some transgenic crops, derived from a bacterial gene.

**back-cross breeding** — Adding a useful gene to a good existing variety by crossing with a plant having that gene and then repeatedly backcrossing to the existing variety while selecting for the desired trait (or, if known, the desired gene).

**bet-hedging** — Sacrificing average performance (investment income, crop yield, and so on) in exchange for lower variability over years.

**biological control** — Use of live organisms (such as predatory insects) to control pests.

**biotechnology** — genetic improvement of crops or livestock by means other than traditional breeding, such as adding genes modified in a lab or taken from another species. See also transgenic.

**C4 photosynthesis** — A form of photosynthesis that increases water-use efficiency (WUE) and eliminates photorespiration by concentrating $CO_2$ in special compartments within leaves. See also photosynthesis.

**carbon dioxide ($CO_2$)** — The atmospheric gas that plants take up via photosynthesis. $CO_2$ levels in the atmosphere are increasing. See also photosynthesis.

**carrying capacity** — The number of animals or plants that can survive or reproduce with the resources available in some environment. A population may exceed carrying capacity for a while, often causing damage that reduces long-term carrying capacity.

**cheating** — By analogy with humans, but with no implication of cognition, investing less than others in a two-species interaction that is potentially beneficial to both.

**chloroplasts** — Photosynthetic organelles within plant cells; distant descendants of photosynthetic symbionts.

**chromosome** — A DNA molecule, typically containing thousands of genes, plus associated proteins.

**constitutive** — Always turned on or always present, such as defensive toxins some plants make even when nothing is eating them.

**crop rotation** — See rotation.

**cue** — Information (perhaps an odor) useful to individuals detecting it, but not necessarily produced because of benefits to sender or receiver.

**dentrification** — Conversion by bacteria of inorganic soil nitrogen in harmless nitrogen gas or other nitrogen-containing gases.

**deoxyribonucleic acid (DNA)** — Millions of nucleotides strung together, containing the information needed to build and operate the bodies of animals and plants, and transmitting that information from one generation to the next.

**development** — A qualitative change in an individual, such as flowering. Contrast with growth, an increase in size, and evolution, a change in the frequency of alleles in a population.

**domatia** — Shelters, typically on leaves, that some plants provide to insects that defend them from pests or pathogens.

**drought tolerance** — The ability to survive, but not necessarily grow, under drought. Contrast with water-use efficiency (WUE).

**ecosystem** — A natural or an agricultural landscape and the species that live there.

**ecosystem services** — Economic benefits to society from ecosystems (often focusing on natural ecosystems), such as water purification and flood control by natural wetlands, pollination of crops by wild species, production of oxygen by photosynthesis, and so on.

**edge effect** — Unusual conditions at the edge of a field. Edge effects may extend throughout a plot, if the plot is small enough.

**efficiency** — Not a general-purpose synonym for *goodness*, but rather the ratio of some desired product or rate to the amount or rate of resource use. Examples are water-use efficiency (WUE), solar-radiation-use efficiency of photosynthesis, and nitrogen-use efficiency.

**endophyte** — A fungus that lives inside a plant without causing harm, and sometimes providing benefits. Compare with pathogen.

**enzyme** — A protein that speeds a chemical reaction.

**evolution** — A small or large change over time (typically, over generations) in the relative frequency of different alleles in a population or species.

**evolutionarily stable strategy** — A strategy (reproducing earlier versus later, for example) that, if in use by most members of a population, cannot be invaded by any alternative strategy.

**feedlot** — A facility where large numbers of animals raised for milk or meat are confined and fed grain or hay brought from some distance, rather than being allowed to graze in a pasture.

**fitness** — The evolutionary success of an allele (increase over time, relative to alternative alleles), or the relative success of an individual in contributing alleles to future generations.

**gene** — A section of DNA coding for a particular function. Examples can include the DNA coding for a particular protein, or a short section of DNA involved in the control of other genes.

**gene expression** — The impact of a given gene, such as the amount of an enzyme (coded for by a given gene) that actually gets made.

**gene flow** — Movement of genes between populations, via migration of animals with those genes, pollen blown between populations that are usually isolated, and so on.

**gene pool** — The sum of all the genomes in a population.

**genetic drift** — Evolution due to chance, such as random effects of conditions on the fitness of individuals (for example, a poorly adapted seed happening to land in an unusually favorable spot). Genetic drift is mainly important in small populations, where the fitness of a few individuals can have a major effect.

**genetic engineer** — A practitioner of biotechnology.

**genomics** — Sequencing whole genomes, rather than focusing on particular hypotheses or genes of interest.

**genotype or genome** — The totality of an individual's genes. See also phenotype.

**greenhouse gas** — A gas that tends to warm the planet by absorbing outgoing radiant heat and returning some of it to the earth's surface; examples include carbon dioxide and methane.

**green manure** — A crop grown mainly to add nitrogen and organic matter to soil.

**Green Revolution** — The most dramatic increase in crop yields of the last century, mostly due to genetic improvements in rice and wheat and increased use of nitrogen fertilizer.

**group selection** — Selection based on group-level traits, often feasible for human-imposed selection, but rare in nature, in contrast to kin selection.

**harvest index** — Ratio of harvested material (for example, grain) to total crop weight; in practice, often excluding roots.

**herbicide** — A chemical used to kill weeds. Some herbicides may also damage certain crops, if a crop susceptible to an herbicide is grown too soon after it is applied, or if wind carries sprays onto a nearby crop.

**herbivores** — Animals, including insects, that eat plants.

**high dose/refuge strategy** — A way of slowing the evolution of pesticide resistance, using a high pesticide dose and gene flow from refuges full of susceptible pests.

**homeostatic mechanism** — A feedback control that maintains stability, like those that regulate body temperature or blood sugar levels in humans, or the less-consistent regulation of natural populations by food supply or predators.

**horizontal gene transfer** — Movement of genes among individuals, often bacteria, independent of reproduction. A bacterium may inherit most of its genes from its parent but acquire a few from other bacteria.

**hypersensitive reaction** — A plant defense against fungal attack in which plant cells infected by the fungus die to prevent the spread of the fungus to other cells of the same plant.

**hypothesis** — An explanatory statement specific enough to make predictions that, if they prove false, require rejection or modification of the hypothesis.

**inducible defense** — A response (such as production of a protective toxin) triggered by pest attack. See also constitutive.

**inherited variation** — Genetic differences within a population; the raw material for natural selection.

**insecticide** — A synthetic or natural chemical that kills insects, preferably pests.

**intensity of selection** — The extent to which individuals with different alleles vary in fitness. If an antibiotic kills 0 percent or 100 percent of bacteria, regardless of genetic differences, the intensity of selection is zero.

**intercropping** — See polyculture and rotation.

**kilograms per hectare (kg/ha)** — A useful measure for grain yield, fertilizer inputs, and so on, roughly equal to pounds per acre. There are 10,000 square meters in a hectare, so, for example, one kilogram (about 2 pounds) of compost per square meter is equivalent to 10,000 kg/ha or 10 metric tons per hectare. That is also a moderate grain yield for irrigated and fertilized corn.

**kin selection** — Natural selection operating via beneficial effects of an allele on the fitness of others with the same allele, usually inherited from the same ancestor.

**last common ancestor** — The most-recent ancestor shared by two or more species (or individuals—for example, the last common ancestors of two first cousins are their grandparents). If some lineages have evolved much less than others, then a species that still exists may resemble an extinct common ancestor—for example, wild sunflowers may still resemble the common ancestor of wild and cultivated sunflowers.

**leaching** — Downward movement through soil, with moving water, of dissolved materials such as agricultural chemicals, eventually polluting aquifers.

**leaf-area index (LAI)** — The ratio of leaf area to land area.

**legume** — A plant family that includes beans, peas, alfalfa, lupines, peanuts, and many other important species, many of which obtain nitrogen by hosting rhizobia in bumps on their roots, known as nodules.

**Long-Term Research on Agricultural Systems (LTRAS)** — A field experiment at the University of California, Davis, comparing the sustainability of ten different combinations of crops and farming methods, over a planned duration of 100 years.

**manipulation** — An activity that changes the behavior of another, harming the other but benefiting the manipulator, as when a spider uses moth sex hormone to lure male moths to be eaten.

**mast seeding** — Synchronized seed production by neighboring plants of the same species, such that seed-eating pests cannot eat them all.

**metabolomics** — Automated analysis of a wide range of molecules other than DNA and protein.

**mitochondria** — Organelles within plant and animal cells that generate energy via respiration. Mitochondria are distant descendants of symbiotic bacteria; they are typically inherited only through the female line.

**monoculture** — Growing only one species (often, only one variety) of crop per field. See also polyculture.

**mutation** — A change in DNA, potentially inherited by offspring.

**mutualism** — An interaction between species that is consistently, or at least usually, beneficial to both. Not all symbioses are mutualistic, and pollination mutualisms, for example, are not intimate enough to be symbioses.

**mycorrhizal fungi** — Symbiotic fungi that connect to plant roots, typically supplying the plant with phosphorus from the soil, while receiving photosynthate.

**natural selection** — Changes in the relative frequency of alternative alleles in a population, due to their effects on survival and reproduction in a given environment.

**nitrogen harvest index** — The fraction of total plant nitrogen that ends up in the harvested portion, such as in grain.

**nonrenewable resources** — Resources, like fossil fuels or high-phosphorus ores, that are replenished so slowly by natural processes that current use decreases future availability.

**nucleotide** — Any of the four "letters" in the DNA "alphabet." A triplet of three nucleotides can code for an amino acid in a protein, but nucleotide sequence can affect phenotypes in many different ways.

**oligoculture** — Reliance on only a few crop species.

**organic farming** — As opposed to conventional farming. Following certain rules, which vary among countries but usually limit the use of synthetic pesticides, fertilizers not derived from natural sources, and transgenic crops.

**pathogen** — A microbe that can cause disease.

**pesticide** — Chemical used to kill pests, such as insects that attack crop plants.

**phenotype** — The sum of an individual's traits, resulting from interactions of genotype and environment.

**phenotypic plasticity (acclimation)** — Differences in phenotype formed by a single genotype, in response to conditions.

**photorespiration** — A source of inefficiency in photosynthesis, resulting from interaction of rubisco with oxygen rather than $CO_2$.

**photosynthesis** — Uptake by plants of carbon dioxide from the atmosphere and conversion to sugars or other carbon-containing compounds, collectively known as photosynthate, using energy in sunlight and typically losing (transpiring) water in the process.

**phylogeny** — The family tree of a group of species, typically including ancestors that are now extinct, but whose traits may be inferred by ancestral-state reconstruction.

**polyculture (intercropping)** — Growing two or more crop species, mixed together, at the same time.

**population** — A subset of a species, somewhat isolated reproductively.

**proteoid roots** — Short-lived clusters of roots, found on only a few plant species, that pump out enzymes and so on to extract phosphorus from soil or from decaying plant materials.

**proteomics** — Automated analysis of multiple proteins.

**quorum sensing** — Bacterial responses to their own population density, based on chemical signals or cues.

**recombination** — A change in the association between genes, as when genes from each parent end up together on a chromosome.

**refuging** — Concentrating resources from a large area into a smaller area.

**rhizobia** — Soil bacteria best-known for spending time in the root nodules of legumes (beans, clover, and so on), where they take up nitrogen gas from air-spaces in the soil and convert it to forms plants can use to make protein.

**ribonucleic acid (RNA)** — Like DNA, RNA consists of long strands of nucleotides. RNA has many functions, with more discovered each year:

messenger RNA carries DNA-sequence information to ribosomes (made mostly of ribosomal RNA), where it directs protein synthesis with help from transfer RNA.

**rotation** — Growing two or more crop species sequentially, typically in successive years.

**Rothamsted (England)** — Location of the world's longest-running experiments on agricultural sustainability (or anything else), started in 1843.

**rubisco** — The enzyme responsible for $CO_2$ uptake during photosynthesis.

**sanctions** — A response to cheating that reduces the fitness of the cheater.

**signal** — A message that benefits both sender and receiver.

**soil organic matter** — A mixture of carbon-containing molecules providing numerous benefits, including improved water storage.

**species** — A group whose members can interbreed, producing fertile offspring.

**stomata** — Pores in the surface of leaves that let carbon dioxide in and water vapor out. Plants can close stomata to limit water loss, although that also tends to limit photosynthesis.

**sustainable** — Possible to continue indefinitely; often misused to describe specific practices having various desirable features but no proven link to sustainability. We can speculate about which practices are most sustainable, but proof depends on data from farms or research plots managed consistently for decades. See also Rothamsted and Long-Term Research on Agricultural Systems.

**symbiosis** — A close, long-lasting interaction between individuals of different species. Compare with mutualism.

**theory** — A collection of facts and principles with broad explanatory and predictive power; examples include the germ theory of infectious disease and the theory of evolution by natural selection.

**tradeoff** — Any limitation on the ability to optimize two goals simultaneously.

**tragedy of the commons** — Broadly, any situation where similar individuals, each following the strategy that maximizes individual benefit, thereby reduce their collective benefits.

**transfer RNA (tRNA)** — An RNA molecule that helps translate the genetic code, during protein synthesis, by connecting a DNA triplet to its corresponding amino acid.

**transgenic** — Containing genes from another species, such as crops that make an insect-killing protein derived from a bacterial gene.

**transpiration** — Evaporation of water from leaves via stomata; the major way that crops consume water.

**trichome** — A hair-shaped gland on a leaf, often producing a chemical that protects the leaf from insect pests.

**water-use efficiency (WUE)** — The ratio of food produced to water used. On a per-leaf basis, WUE is the ratio of photosynthesis to transpiration.

**yield (land-use efficiency)** — The ratio of food harvested (per year or per crop) to land area used to grow the crop.

**yield gap** — The difference between yield potential and average yields in the same country.

**yield monitor** — A device built into harvesting equipment that automatically measures yields for each small area within the field.

**yield potential** — The yield achievable by the best farmers in a region, with excellent control of weeds, pests, and disease. By this definition, a disease-resistant crop may have greater yield, but not greater yield potential.

# References

1 Coyne, J. A., 2009. *Why Evolution Is True*. Viking, New York.

2 Dawkins, R., 2009. *The Greatest Show on Earth: The Evidence for Evolution*. Free Press, New York.

3 Zimmer, C., 2009. *The Tangled Bank: An Introduction to Evolution*. Roberts and Company, Greenwood Village, CO.

4 Tcherkez, G.G.B., Farquhar, G. D., and Andrews, T. J., 2006. Despite slow catalysis and confused substrate specificity, all ribulose bisphosphate carboxylases may be nearly perfectly optimized. *Proc. Natl. Acad. Sci. USA* 103, 7246–7251.

5 Dawkins, R., 1982. *The Extended Phenotype*. Oxford University Press, Oxford, UK.

6 *The Holy Bible, Old and New Testaments in the King James Version*, 1972. Thomas Nelson Publishers, Nashville, TN.

7 Brown, L. R., 2011. *World on the Edge: How to Prevent Environmental and Economic Collapse*. W.W. Norton, New York.

8 Cassman, K. G., 1999. Ecological intensification of cereal production systems: yield potential, soil quality, and precision agriculture. *Proc. Natl. Acad. Sci. USA* 96, 5952–5959.

9 Condon, A. G., Richards, R. A., Rebetzke, G. J., and Farquhar, G. D., 2004. Breeding for high water-use efficiency. *J. Exp. Bot.* 55, 2447–2460.

10 Foley, J. A., Monfreda, C., Ramankutty, N., and Zaks, D., 2007. Our share of the planetary pie. *Proc. Natl. Acad. Sci. USA* 104, 12585–12586.

11 Loomis, R. S., and Connor, D. J., 1992. *Crop Ecology: Productivity and Management in Agricultural Systems*. Cambridge University Press, Cambridge, UK.

12 Conway, G., 1998. *The Doubly Green Revolution: Food for All in the Twenty-First Century*. Comstock Publishing Associates, Ithaca, NY.

13 Fedoroff, N. V., Battisti, D. S., Beachy, R. N., Cooper, P. J., et al., 2010. Radically rethinking agriculture for the 21st century. *Science* 327, 833–834.

14 Jackson, W., and Piper, J., 1989. The necessary marriage between ecology and agriculture. *Ecology* 70, 1591–1593.

15 Soule, J. D., and Piper, J. K., 1992. *Farming in Nature's Image. An Ecological Approach to Agriculture*. Island Press, Corelo, CA.

16 Connor, D. J., Loomis, R. S., and Cassman, K. G., 2011. *Crop Ecology: Productivity and Management in Agricultural Systems*. Cambridge University Press, Cambridge, UK.

17 Stowe, M. K., Turlings, T.C.J., Loughrin, J. H., Lewis, W. J., et al., 1995. The chemistry of eavesdropping, alarm, and deceit. *Proc. Natl. Acad. Sci. USA* 92, 23–28.

18 Denison, W. C., 1973. Life in tall trees. *Sci. Am.* 228, 74–80.

[19] Mueller, U. G., Rehner, S. A., and Schultz, T. R., 1998. The evolution of agriculture in ants. *Science* 281, 2034–2038.

[20] Kingsolver, B., 2007. *Animal, Vegetable, Miracle: A Year of Food Life*. Harper Collins, New York.

[21] Mueller, U. G., Gerardo, N. M., Aanen, D. K., Six, D. L., et al., 2005. The evolution of agriculture in insects. *Annu. Rev. Ecol. Evol. Syst.* 36, 563–595.

[22] Reynolds, H. T., and Currie, C. R., 2004. Pathogenicity of *Escovopsis weberi*: the parasite of the attine ant-microbe symbiosis directly consumes the ant-cultivated fungus. *Mycologia* 96, 955–959.

[23] Currie, C. R., Scott, J. A., Summerbell, R. C., and Malloch, D., 1999. Fungus-growing ants use antibiotic-producing bacteria to control garden parasites. *Nature* 398, 701–704.

[24] Pennisi, E., 2003. Drafting a tree. *Science* 300, 1694.

[25] Trewavas, T., 2008. Redefining "natural" in agriculture. *PLoS Biol.* 6, 1618–1620.

[26] Darwin, C. R., 1859. *On the Origin of Species by Means of Natural Selection, Or the Preservation of Favoured Races in the Struggle for Life*. John Murray, London.

[27] Zhu, X. G., Portis, A. R., and Long, S. P., 2004. Would transformation of C3 crop plants with foreign rubisco increase productivity? A computational analysis extrapolating from kinetic properties to canopy photosynthesis. *Plant Cell Environ.* 27, 155–165.

[28] Denison, R. F., Kiers, E. T., and West, S. A., 2003. Darwinian agriculture: when can humans find solutions beyond the reach of natural selection? *Q. Rev. Biol.* 78, 145–168.

[29] Denison, R. F., 2011. Past evolutionary tradeoffs represent opportunities for crop genetic improvement and increased human lifespan. *Evol. Appl.* 4, 216–224.

[30] Denison, R. F., Bryant, D. C., Kearney, T. E., 2004. Crop yields over the first nine years of LTRAS, a long-term comparison of field crop systems in a Mediterranean climate. *Field Crops Res.* 86, 267–277.

[31] BBC News, 2011. Food prices hit new record highs, says UN food agency. 3 March 2011.

[32] Malthus, T. R., 1807. *An Essay on the Principle of Population*. Printed for J. Johnson by T. Bensley, London.

[33] Evans, L. T., 1993. *Crop Evolution, Adaptation, and Yield*. Cambridge University Press, Cambridge, UK.

[34] Cohen, J. E., 1995. *How Many People Can the Earth Support?* W.W. Norton, New York.

[35] McWilliams, J. E., 2007. Food that travels well. *New York Times*, 6 August 2007.

[36] Weber, C. L., and Matthews, H. S., 2008. Food-miles and the relative climate impacts of food choices in the United States. *Environ. Sci. Technol.* 42, 3508–3513.

[37] Denison, R. F., and Kiers, E. T., 2005. Sustainable crop nutrition: constraints and opportunities. In *Plant Nutritional Genomics*, M. Broadley, ed., pp. 242–264. Blackwell Publishing, Oxford, UK.

[38] Stanhill, G., 1990. The comparative productivity of organic agriculture. *Agric. Ecosyst. Environ.* 30, 1–26.

[39] MacDonald, G. K., Bennett, E. M., Potter, P. A., and Ramankutty, N., 2011. Agronomic phosphorus imbalances across the world's croplands. *Proc. Natl. Acad. Sci. USA* 108, 3086–3091.

[40] Liu, J., You, L., Amini, M., Obersteiner, M., et al., 2010. A high-resolution assessment on global nitrogen flows in cropland. *Proc. Natl. Acad. Sci. USA* 107, 8035–8040.

[41] Muller-Landau, H. C., and Wright, S. J., 2006. The future of tropical forest species. *Biotropica* 38, 287–301.

[42] Jacobsen, T., and Adams, R. M., 1958. Salt and silt in ancient Mesopotamian agriculture. *Science* 128, 1251–1258.

[43] Hasegawa, H., Labavitch, J. M., McGuire, A. M., Bryant, D. C., et al., 1999. Testing CERES model predictions of N release from legume cover crop residue. *Field Crops Res.* 63, 255–267.

[44] Johnston, A. E., and Powlson, D. S., 1994. The setting-up, conduct and applicability of long-term, continuing field experiments in agricultural research. In *Soil Resilience and Sustainable Land Use*, D. J. Greenland and I. Szabolcs, eds., pp. 395–421. CAB International, Wallingford, UK.

[45] Jenkinson, D. S., 1991. The Rothamsted long-term experiments: Are they still of use? *Agron. J.* 83, 2–10.

[46] Bearchell, S. J., Fraaije, B. A., Shaw, M. W., and Fitt, B.D.L., 2005. Wheat archive links long-term fungal pathogen population dynamics to air pollution. *Proc. Natl. Acad. Sci. USA* 102, 5438–5442.

[47] Powlson, D. S., and Johnston, A. E., 1994. Long-term field experiments: their importance in understanding sustainable land use. In *Soil Resilience and Sustainable Land Use*, D. J. Greenland and I. Szabolcs, eds., pp. 367–394. CAB International, Wallingford, UK.

[48] Fisher, R. A., 1930. *The Genetical Theory of Natural Selection.* Oxford University Press, Oxford, UK.

[49] Fisher, R. A., 1930. *Statistical Methods for Research Workers.* Oliver and Boyd, Edinburgh, UK.

[50] Okano, Y., Hristova, K. R., Leutenegger, C. M., Jackson, L. E., et al., 2004. Application of real-time PCR to study effects of ammonium on population size of ammonia-oxidizing bacteria in soil. *Appl. Environ. Microbiol.* 70, 1008–1016.

[51] Adams, D., 1979. *The Hitch-Hiker's Guide to the Galaxy.* Pan Books, London.

[52] Nelson, T., 2007. Prairie project may reshape the renewable fuel landscape. *Solutions* 1, 14–16.

[53] Cordell, D., Drangert, J., and White, S., 2009. The story of phosphorus: global food security and food for thought. *Global Environ. Change* 19, 292–305.

[54] Turgeon, S. C., and Creaser, R. A., 2008. Cretaceous oceanic anoxic event 2 triggered by a massive magmatic episode. *Nature* 454, 323–326.

[55] Harris, A., 2008. What Spaceguard did. *Nature* 453, 1178–1179.

[56] Kaiser, J., 2005. Resurrected influenza virus yields secrets of deadly 1918 pandemic. *Science* 310, 28–29.

[57] Constanza, R., D'Arge, R., de Groot, R., Farber, S., et al., 1997. The value of the world's ecosystem services and natural capital. *Nature* 387, 253–260.

[58] Alexander, R. B., Smith, R. A., Schwarz, G. E., Boyer, E. W., et al., 2008. Differences in phosphorus and nitrogen delivery to the Gulf of Mexico from the Mississippi River basin. *Environ. Sci. Technol.* 42, 822–830.

[59] Elphick, C. S., 2000. Functional equivalency between rice fields and seminatural wetland habitats. *Conserv. Biol.* 14, 181–191.

[60] Tangley, L., 1996. The case of the missing migrants. *Science* 274, 1299–1300.

[61] Waggoner, P. E., 1994. *How Much Land Can Ten Billion People Spare for Nature?* Council for Agricultural Science and Technology, Ames, IA.

[62] Green, R. E., Cornell, S. J., Scharlemann, J.P.W., and Balmford, A., 2004. Farming and the fate of wild nature. *Science* 307, 550–555.

[63] Matson, P. A., and Vitousek, P. M., 2006. Agricultural intensification: will land spared from farming be land spared for nature? *Conserv. Biol.* 20, 709–710.

[64] Kinraide, T. B., and Denison, R. F., 2003. Strong inference, the way of science. *Am. Biol. Teach.* 65, 419–424.

[65] Johannsen, W., 1911. The genotype conception of heredity. *Am. Nat.* 45, 129–159.

[66] Stanford, E. H., Laude, H. M., and Enloe, J. A., 1960. Effect of harvest dates and location in the genetic composition of the Syn1 generation of Pilgrim Ladino clover. *Agron. J.* 52, 149–152.

[67] Rainey, P. B., and Travisano, M., 1998. Adaptive radiation in a heterogeneous environment. *Nature* 394, 69–72.

[68] Whitton, J., Wolf, D. E., Arias, D. M., Snow, A. A., et al., 1997. The persistence of cultivar alleles in wild populations of sunflowers five generations after hybridization. *Theor. Appl. Genet.* 95, 33–40.

[69] Kondrashov, A. S., 2002. Direct estimates of human per nucleotide mutation rates at 20 loci causing Mendelian diseases. *Hum. Mutat.* 21, 12–27.

[70] Koch, M. A., Haubold, B., and Mitchell-Olds, T., 2000. Comparative evolutionary analysis of chalcone synthase and alcohol dehydrogenase loci in *Arabidopsis, Arabis,* and related genera (Brassicaceae). *Mol. Biol. Evol.* 17, 1483–1498.

[71] Calcott, B., 2008. Assessing the fitness landscape revolution. *Biol. Philos.* 23, 639–657.

[72] Weinreich, D. M., Delaney, N. F., DePristo, M. A., and Hartl, D. L., 2006. Darwinian evolution can follow only very few mutational paths to fitter proteins. *Science* 312, 111–114.

[73] Poelwijk, F. J., Kiviet, D. J., Weinreich, D. M., and Tans, S. J., 2007. Empirical fitness landscapes reveal accessible evolutionary paths. *Nature* 445, 383–386.

[74] Meer, M. V., Kondrashov, A. S., Artzy-Randrup, Y., and Kondrashov, F. A., 2010. Compensatory evolution in mitochondrial tRNAs navigates valleys of low fitness. *Nature* 464, 279–282.

[75] Margulis, L., and Bermudes, D., 1985. Symbiosis as a mechanism of evolution: status of cell symbiosis theory. *Symbiosis* 1, 101–124.

[76] Maynard Smith, J., and Szathmáry, E., 1995. *The Major Transitions in Evolution.* Oxford University Press, NY.

[77] Huelsenbeck, J. P., and Bollback, J. P., 2001. Empirical and hierarchical Bayesian estimation of ancestral states. *Syst. Biol.* 50, 351–366.

[78] Dalrymple, G. B., 2004. *Ancient Earth, Ancient Skies: The Age of Earth and Its Cosmic Surroundings.* Stanford University Press, Stanford, CA.

[79] Wells, J. W., 1963. Coral growth and geochronometry. *Nature* 197, 948–950.

[80] Nilsson, D. E., and Pelger, S., 1994. A pessimistic estimate of the time required for an eye to evolve. *Proc. Roy. Soc. Lond. B* 256, 53–58.

[81] Berenbaum, M. R., and Zangerl, A. R., 1998. Chemical phenotype matching between a plant and its insect herbivore. *Proc. Natl. Acad. Sci. USA* 95, 13743–13748.

[82] Frank, R. H., and Bernanke, B. S., 2009. *Principles of Microeconomics.* McGraw-Hill, Boston.

[83] Duncan, W. G., Williams, W. A., and Loomis, R. S., 1967. Tassels and productivity of maize. *Crop Sci.* 7, 37–39.

[84] Duvick, D. N., and Cassman, K. G., 1999. Post-green-revolution trends in yield potential of temperate maize in the north-central United States. *Crop Sci.* 39, 1622–1630.

[85] Kakes, P., 1989. An analysis of the costs and benefits of the cyanogenic system in *Trifolium repens* L. *Theor. Appl. Genet.* 77, 111–118.

[86] Agrawal, A. A., and Karban, R., 1999. Why induced defenses may be favored over constitutive strategies in plants. In *The Ecology and Evolution of Inducible Defenses*, R. Tollrian and C. D. Harvell, eds., pp. 45–61. Princeton University Press, Princeton, NJ.

[87] Engelberth, J., Alborn, H. T., Schmelz, E. A., and Tumlinson, J. H., 2004. Airborne signals prime plants against insect herbivore attack. *Proc. Natl. Acad. Sci. USA* 101, 1781–1785.

[88] Wolpert, T. J., Dunkle, L. D., and Ciuffetti, L. M., 2002. Host-selective toxins and avirulence determinants: what's in a name? *Annu. Rev. Phytopathol.* 40, 251–285.

[89] Hermanns, M., Slusarenko, A. J., and Schlaich, N. L., 2003. Organ-specificity in a plant disease is determined independently of R gene signaling. *Mol. Plant Microbe Interact.* 16, 752–759.

[90] Williams, G. C., 1957. Pleiotropy, natural selection, and the evolution of senescence. *Evolution* 11, 398–411.

[91] Hamilton, W. D., 1966. The moulding of senescence by natural selection. *J. Theor. Biol.* 12, 12–45.

[92] Blomquist, G. E., 2009. Trade-off between age of first reproduction and survival in a female primate. *Biol. Lett.* 5, 339–342.

[93] Ratcliff, W. C., Hawthorne, P., Travisano, M., and Denison, R. F., 2009. When stress predicts a shrinking gene pool, trading early reproduction for longevity can increase fitness, even with lower fecundity. *PLoS One* 4, e6055.

[94] Behboodi, B. S., 2005. Ecological distribution study of wild pistachios for selection of rootstock. *Options Méditerranéennes* 63, 61–67.

[95] Stevenson, M. T., and Shackel, K. A., 1998. Alternate bearing in pistachio as a masting phenomenon: construction cost of reproduction versus vegetative growth and storage. *J. Am. Soc. Hort. Sci.* 123, 1069–1075.

[96] Standage, T., 1998. *The Victorian Internet.* Berkley Books, New York.

[97] Lobell, D. B., Cassman, K. G., and Field, C. B., 2009. Crop yield gaps: their importance, magnitudes, and causes. *Annu. Rev. Environ. Resour.* 34, 179–204.

[98] Evans, L. T., and Fischer, R. A., 1999. Yield potential: its definition, measurement, and significance. *Crop Sci.* 39, 1544–1551.

[99] Karban, R., Huntzinger, M., and McCall, A. C., 2004. The specificity of eavesdropping on sagebrush by other plants. *Ecology* 85, 1846–1852.

[100] Bell, M. A., Fischer, R. A., Byerlee, D., and Sayre, K., 1995. Genetic and agronomic contributions to yield gains: a case study for wheat. *Field Crops Res.* 44, 55–65.

[101] Peng, S., Cassman, K. G., Virmani, S. S., Sheehy, J., et al., 1999. Yield potential trends of tropical rice since the release of IR8 and the challenge of increasing rice yield potential. *Crop Sci.* 39, 1552–1559.

[102] Kellogg, E. A., 1999. Phylogenetic aspects of the evolution of C4 photosynthesis. In *C4 Plant Biology*, R. F. Sage and R. K. Monson, eds., pp. 411–444. Academic Press, San Diego, CA.

[103] Hartsock, T. L., and Nobel, P. S., 1976. Watering converts a CAM plant to daytime $CO_2$ uptake. *Nature* 262, 574–576.

[104] Hibberd, J. M., Sheehy, J. E., and Langdale, J. A., 2008. Using C4 photosynthesis to increase the yield of rice—rationale and feasibility. *Curr. Opin. Plant Biol.* 11, 228–231.

[105] Zelitch, I., 1975. Improving the efficiency of photosynthesis. *Science* 188, 626–633.

[106] Somerville, C. R., and Ogren, W. L., 1982. Genetic modification of photorespiration. *Trends Biochem. Sci.* 7, 171–174.

[107] Mann, C. C., 1999. Genetic engineers aim to soup up crop photosynthesis. *Science* 283, 314–316.

[108] Ku, M.S.B., Cho, D., Ranade, U., Hsu, T. P., et al., 2000. Photosynthetic performance of transgenic rice plants overexpressing maize C4 photosynthesis enzymes. In *Redesigning Rice Photosynthesis to Increase Yield*, J. E. Sheehy, P. L. Mitchell, and B. Hardy, eds., pp. 193–204. International Rice Research Institute and Elsevier, Amsterdam.

[109] Matsuoka, M., Fukayama, H., Tsuchida, H., Nomura, M., et al., 2000. How to express some C4 photosynthesis genes at high levels in rice. In *Redesigning Rice Photosynthesis to Increase Yield*, J. E. Sheehy, P. L. Mitchell, and B. Hardy, eds., pp. 167–175. International Rice Research Institute and Elsevier, Amsterdam.

[110] Long, S. P., Zhu, X. G., Naidu, S. L., and Ort, D. R., 2006. Can improvement in photosynthesis increase crop yields? *Plant Cell Environ.* 29, 315–330.

[111] Pendleton, J. W., Smith, G. E., Winter, S. R., and Johnston, T. J., 1968. Field investigations of the relationships of leaf angle in corn *(Zea mays* L.) to grain yield and apparent photosynthesis. *Agron. J.* 60, 422–424.

[112] Donald, C. M., 1968. The breeding of crop ideotypes. *Euphytica* 17, 385–403.

[113] Marris, E., 2008. More crop per drop. *Nature* 452, 273–277.

[114] Nelson, D. E., Repetti, P. P., Adams, T. R., Creelman, R. A., et al., 2007. Plant nuclear factor Y (NF-Y) B subunits confer drought tolerance and lead to improved corn yields on water-limited acres. *Proc. Natl. Acad. Sci. USA* 104, 16450–16455.

[115] Sinclair, T. R., Purcell, L. C., and Sneller, C. H., 2004. Crop transformation and the challenge to increase yield potential. *Trends Plant Sci.* 9, 70–75.

[116] Gurian-Sherman, D., 2009. *Failure to Yield: Evaluating the Performance of Genetically Engineered Crops*. Union of Concerned Scientists, Cambridge, MA.

[117] Paris, M., 2001. Vapornomics. *Nat. Biotechnol.* 19, 301.

[118] Jacob, F., 1977. Evolution and tinkering. *Science* 196, 1161–1166.

[119] Kebeish, R., Niessen, M., Thiruveedhi, K., Bari, R., et al., 2007. Chloroplastic photorespiratory bypass increases photosynthesis and biomass production in *Arabidopsis thaliana*. *Nat. Biotechnol.* 25, 593–599.

[120] Daniell, H., and Gepts, P., 2004. Alternative pharma crops. In *A Growing Concern: Protecting the Food Supply in an Era of Pharmaceutical and Industrial Crops*, D. Andow, ed., pp. 102–108. Union of Concerned Scientists, Cambridge, MA.

[121] Siritunga, D., and Sayre, R., 2004. Engineering cyanogen synthesis and turnover in cassava (*Manihot esculenta*). *Plant Mol. Biol.* 56, 661–669.

[122] Ellstrand, N. C., 2001. When transgenes wander, should we worry? *Plant Physiol.* 125, 1543–1545.

[123] Arriola, P. E., and Ellstrand, N. C., 1996. Crop-to-weed gene flow in the genus *Sorghum* (Poaceae): spontaneous intraspecific hybridization between johnsongrass, *Sorghum halepense*, and crop sorghum, *S. bicolor*. *Am. J. Bot.* 83, 1153–1160.

[124] Perez-Jones, A., Park, K. W., Polge, N., Colquhoun, J., et al., 2007. Investigating the mechanisms of glyphosate resistance in *Lolium multiflorum*. *Planta* 226, 395–404.

[125] Pratley, J., Baines, P., Eberbach, P., Incerti, M., et al., 1996. Glyphosate resistance in annual ryegrass. *Proceedings, Annual Conference of the Grassland Society of New South Wales* 11, 122.

[126] VanGessel, M. J., 2001. Glyphosate-resistant horseweed from Delaware. *Weed Sci.* 49, 703–705.

[127] Snow, A. A., and Palma, P. M., 1997. Commercialization of transgenic plants: potential ecological risks. *Bioscience* 47, 86–96.

[128] Perry, J. N., Firbank, L. G., Champion, G. T., Clark, S. J., et al., 2004. Ban on triazine herbicides likely to reduce but not negate relative benefits of GMHT maize cropping. *Nature* 428, 313–316.

[129] Zheutlin, P., 2008. Annie Londonderry's extraordinary ride. *Women in Judaism: A Multidisciplinary Journal* 5, 1–2.

[130] Abbott, A., 2008. Swiss "dignity" law is threat to plant biology. *Nature* 452, 919.

[131] Gutierrez, A. P., and Daxl, R., 1984. Economic thresholds for cotton pests in Nicaragua: ecological and evolutionary perspectives. In *Pest and Pathogen Control, Strategic, Tactical, and Policy Models*, G. R. Conway, ed., pp. 184–205. John Wiley and Sons, New York.

[132] Altieri, M. A., 1987. *Agroecology: The Scientific Basis of Alternative Agriculture*. Westview Press, Boulder, CO.

[133] Lefroy, E. C., Hobbs, R. J., O'Connor, M. H., and Pate, J. S., eds., 1999. *Agriculture as a Mimic of Natural Ecosystems*. Kluwer Academic Publishers, Dordrecht, The Netherlands.

[134] Feynman, R., 1985. *Surely You're Joking, Mr. Feynman!* Vintage, London.

[135] Malakoff, D., 1999. Fighting fire with fire. *Science* 285, 1841–1843.

[136] Gardner, A., and Grafen, A., 2009. Capturing the superorganism: a formal theory of group adaptation. *J. Evol. Biol.* 22, 659–671.

[137] Maynard Smith, J., 1964. Group selection and kin selection. *Nature* 201, 1145–1147.

[138] Levin, B. R., and Kilmer, W. L., 1974. Interdemic selection and the evolution of altruism: a computer simulation study. *Evolution* 28, 527–545.

[139] Leigh, E. G., 1983. When does the good of the group override the advantage of the individual? *Proc. Natl. Acad. Sci. USA* 80, 2985–2989.

[140] Hamilton, W. D., 1963. The evolution of altruistic behavior. *Am. Nat.* 97, 354–356.

[141] Williams, G. W., 1966. *Adaptation and Natural Selection*. Princeton University Press, Princeton, NJ.

[142] Dawkins, R., 1976. *The Selfish Gene*. Oxford University Press, Oxford, UK.

[143] Tatum, L. A., 1971. The southern corn leaf blight epidemic. *Science* 171, 1113–1116.

[144] Maynard Smith, J., 1991. A Darwinian view of symbiosis. In *Symbiosis as a Source of Evolutionary Innovation*, L. Margulis and R. Fester, eds., pp. 26–39. MIT Press, Cambridge, MA.

[145] Dominguez, C. A., 1995. Genetic conflicts of interest in plants. *Trends Ecol. Evol.* 10, 412–416.

[146] McIntosh, R. P., 1998. The myth of community as organism. *Perspect. Biol. Med.* 41, 426–438.

[147] Wood, D., and Lenne, J., 2001. Nature's fields: a neglected model for increasing food production. *Outlook Agric.* 30, 161–170.

[148] Ovington, J. D., Haitkamp, D., and Lawrence, D. B., 1963. Plant biomass and productivity of prairie, savanna, oakwood, and maize field ecosystems in central Minnesota. *Ecology* 44, 52–63.

[149] Brown, H. R., 1943. Growth and seed yields of native prairie plants in various habitats of the mixed-prairie. *Trans. Kans. Acad. Sci.* 46, 87–99.

[150] Bailey, R. C., and Headland, T. N., 1991. The tropical rain forest: is it a productive environment for human foragers? *Hum. Ecol.* 19, 261–285.

[151] Suyamto, H., 1998. Potassium increases cassava yield on alfisol soils. *Better Crops Int.* 12, 12–13.

[152] Feldhamer, G. A., Kilbane, T. P., and Sharp, D. W., 1989. Cumulative effect of winter on acorn yield and deer body weight. *J. Wildl. Manage.* 53, 292–295.

[153] Hutmacher, R. B., Nightingale, H. I., Rolston, D. E., Biggar, J. W., et al., 1994. Growth and yield responses of almond (*Prunus amygdalus*) to trickle irrigation. *Irrig. Sci.* 14, 117–126.

[154] Pastor, J., and Walker, R. D., 2006. Delays in nutrient cycling and plant population oscillations. *Oikos* 112, 698–705.

[155] Moyle, J. B., 1944. Wild rice in Minnesota. *J. Wildl. Manage.* 8, 177–184.

[156] Oelke, E. A., Grava, J., Noetzel, D., Barron, D., et al., 1982. *Wild Rice Production in Minnesota*. University of Minnesota, Saint Paul.

[157] Hachten, H., and Allen, T., 1981. *The Flavor of Wisconsin: An Informal History of Food and Eating in the Badger State*. Wisconsin Historical Society Press, Madison.

[158] Leigh, E. G., and Vermeij, G. J., 2001. Does natural selection organize ecosystems for the maintenance of high productivity and diversity? *Philos. Trans. Roy. Soc. Lond. B* 357, 709–718.

[159] Gotherstrom, A., Anderung, C., Hellborg, L., Elburg, R., et al., 2005. Cattle domestication in the Near East was followed by hybridization with aurochs bulls in Europe. *Proc. Roy. Soc. Lond. B* 272, 2345–2350.

[160] Kumpula, J., 2001. Productivity of the semi-domesticated reindeer (*Rangifer t. tarandus* L.) stock and carrying capacity of pastures in Finland during 1960–1990s, PhD Thesis, University of Oulu, Finland.

[161] Messier, F., 1995. Trophic interactions in two northern wolf-ungulate systems. *Wildlife Res.* 22, 131–146.

[162] Messier, F., Huot, J., le Henaff, D., and Luttich, S., 1988. Demography of the George River caribou herd: evidence of population regulation by forage exploitation and range expansion. *Arctic* 41, 279–287.

[163] Danell, O., 2007. Status, directions, and priorities of reindeer husbandry research in Sweden. *Polar Res.* 19, 111–115.

[164] Boudreau, S., Payette, S., Morneau, C., and Couturier, S., 2003. Recent decline of the George River caribou herd as revealed by tree-ring analysis. *Arct. Antarct. Alp. Res.* 35, 187–195.

[165] Kumpula, J., Colpaert, A., and Nieminen, M., 2000. Condition, potential recovery rate, and productivity of lichen (*Cladonia* spp.) ranges in the Finnish reindeer management area. *Arctic* 53, 152–160.

[166] Kumpula, J., Colpaert, A., and Nieminen, M., 1998. Reproduction and productivity of semidomesticated reindeer in northern Finland. *Can. J. Zool.* 76, 269–277.

[167] Maynard Smith, J., 1978. *The Evolution of Sex*. Cambridge University Press, Cambridge, UK.

[168] Hamilton, W. D., 1967. Extraordinary sex ratios. *Science* 156, 477–488.

[169] Sterner, R. W., 2002. *Ecological Stoichiometry: The Biology of Elements from Molecules to the Biosphere*. Princeton University Press, Princeton, NJ.

[170] Begon, M., Townsend, C., and Harper, J., 2005. *Ecology—from Individuals to Ecosystems*. Blackwell, Oxford, UK.

[171] Risch, S. J., Andow, D., and Altieri, M. A., 1983. Agroecosystem diversity and pest control: data, tentative conclusions, and new research directions. *Environ. Entomol.* 12, 625–629.

[172] Fukai, S., 1993. Intercropping—bases of productivity. *Field Crops Res.* 34, 239–245.

[173] Dohleman, F. G., and Long, S. P., 2009. More productive than maize in the Midwest: how does *Miscanthus* do it? *Plant Physiol.* 150, 2104–2115.

[174] Sinclair, T. R., and de Wit, C. T., 1975. Photosynthate and nitrogen requirements for seed production by various crops. *Science* 189, 565–567.

[175] van Noordwijk, A. J., and de Jong, G., 1986. Acquisition and allocation of resources: their influence on variation in life history tactics. *Am. Nat.* 128, 137–142.

[176] Roff, D. A., and Fairbairn, D. J., 2007. The evolution of trade-offs: where are we? *J. Evol. Biol.* 20, 433–447.

[177] Obeso, J. R., 2002. The costs of reproduction in plants. *New Phytol.* 155, 321–348.

[178] Gillman, M. P., and Crawley, M. J., 1990. The cost of sexual reproduction in ragwort (*Senecio jacobaea* L.). *Funct. Ecol.* 4, 585–589.

[179] Aragon, C. F., Mendez, M., and Escudero, A., 2009. Survival costs of reproduction in a short-lived perennial plant: live hard, die young. *Am. J. Bot.* 96, 904–911.

[180] Agren, J., and Willson, M. F., 1994. Cost of seed production in the perennial herbs *Geranium maculatum* and *G. sylvaticum*: an experimental field study. *Oikos* 70, 35–42.

[181] Suneson, C. A., Sharkawy, A. E., and Hall, W. E., 1963. Progress in 25 years of perennial wheat development. *Crop Sci.* 3, 437–439.

[182] DeHaan, L. R., Van Tassel, D. L., and Cox, T. S., 2007. Perennial grain crops: a synthesis of ecology and plant breeding. *Renewable Agric. Food Systems* 20, 5–14.

[183] DeHaan, L. R., Ehlke, N. J., Sheaffer, C. C., DeHaan, R. L., et al., 2003. Evaluation of diversity among and within accessions of Illinois bundleflower. *Crop Sci.* 43, 1528–1537.

[184] Moffat, A. S., 1996. Higher yielding perennials point the way to new crops. *Science* 274, 1469–1470.

[185] Fischbach, J. A., Peterson, P. R., Sheaffer, C. C., Ehlke, N. J., et al., 2005. Illinois bundleflower forage potential in the upper midwestern USA: I. Yield, regrowth, and persistence. *Agron. J.* 97, 886–894.

[186] Jacob, J. P., 2007. Evaluation of Illinois bundleflower (*Desmanthus illinoensis*) for broiler chicks. *J. Appl. Poult. Res.* 16, 39–44.

[187] Beja-Pereira, A., Luikart, G., England, P. R., Bradley, D. G., et al., 2003. Gene-culture coevolution between cattle milk protein genes and human lactase genes. *Nat. Genet.* 35, 311–313.

[188] Wade, N., 2006. Study detects recent instance of human evolution. *New York Times*, 10 December 2006.

[189] Chaney, D. E., Drinkwater, L. E., and Pettygrove, G. S., 1992. *Organic Soil Amendments and Fertilizers*. University of California, Oakland.

[190] Parr, J. F., Miller, R. H., and Colacicco, D., 1984. Utilization of organic materials for crop production in developed and developing countries. In *Organic Farming: Current Technology and Its Role in a Sustainable Agriculture*, D. F. Bezdicek, J. F. Power, D. R. Keeney, and M. J. Wright, eds., pp. 83–95. American Society of Agronomy, Madison, WI.

[191] Oono, R., and Denison, R. F., 2010. Comparing symbiotic efficiency between swollen versus nonswollen rhizobial bacteroids. *Plant Physiol.* 154, 1541–1548.

[192] Dinkelaker, B., Hengeler, C., and Marschner, H., 1995. Distribution and function of proteoid roots and other root clusters. *Botanica Acta* 108, 183–200.

[193] Hendry, A. P., Kinnison, M. T., Heino, M., Day, T., et al., 2011. Evolutionary principles and their practical application. *Evol. Appl.* 4, 159–183.

[194] Thrall, P. H., Oakeshott, J. G., Fitt, G., Southerton, S., et al., 2011. Evolution in agriculture: the application of evolutionary approaches to the management of biotic interactions in agro-ecosystems. *Evol. Appl.* 4, 200–215.

[195] Lipsitch, M., Bergstrom, C. T., and Levin, B. R., 2000. The epidemiology of antibiotic resistance in hospitals: paradoxes and prescriptions. *Proc. Natl. Acad. Sci. USA* 97, 1938–1943.

[196] Levin, B. R., and Bergstrom, C. T., 2000. Bacteria are different: observations, interpretations, speculations, and opinions about the mechanisms of adaptive evolution in prokaryotes. *Proc. Natl. Acad. Sci. USA* 97, 6981–6985.

[197] Althouse, B. M., Bergstrom, T. C., and Bergstrom, C. T., 2010. A public choice framework for controlling transmissible and evolving diseases. *Proc. Natl. Acad. Sci. USA* 107, 1696–1701.

[198] Rees, W., and Wackernagel, M., 1996. Urban ecological footprints: why cities cannot be sustainable—and why they are a key to sustainability. *Environ. Impact Assess Rev.* 16, 223–248.

[199] Haberl, H., Erb, K., and Krausmann, F., 2001. How to calculate and interpret ecological footprints for long periods of time: the case of Austria 1926–1995. *Ecol. Econ.* 38, 25–45.

[200] Murtaugh, P. A., and Schlax, M. G., 2009. Reproduction and the carbon legacies of individuals. *Global Environ. Change* 19, 14–20.

[201] Diamond, J., 1993. Ten thousand years of solitude. *Discover*, March 1993.

[202] Hamilton, W. J., and Watt, K.E.F., 1970. Refuging. *Annu. Rev. Ecol. Evol. Syst.* 1, 263–286.

[203] Osborne, L. L., and Kovacic, D. A., 1993. Riparian vegetated buffer strips in water-quality restoration and stream management. *Freshwat. Biol.* 29, 243–258.

[204] Tilman, D., Reich, P. B., Knops, J., Wedin, D., et al., 2001. Diversity and productivity in a long-term grassland experiment. *Science* 294, 843–845.

[205] Tilman, D., Wedin, D., and Knops, J., 1996. Productivity and sustainability influenced by biodiversity in grassland ecosystems. *Nature* 379, 718–720.

[206] Fargione, J. E., and Tilman, D., 2005. Diversity decreases invasion via both sampling and complementarity effects. *Ecol. Lett.* 8, 604–611.

[207] HilleRisLambers, J., Harpole, W. S., Schnitzer, S., Tilman, D., et al., 2009. $CO_2$, nitrogen, and diversity differentially affect seed production of prairie plants. *Ecology* 90, 1810–1820.

[208] Gliessman, S. R., and Altieri, M. A., 1982. Polyculture cropping has advantages. *Calif. Agric.* July 1982, 14–16.

[209] Hooper, D. U., and Dukes, J. S., 2004. Overyielding among plant functional groups in a long-term experiment. *Ecol. Lett.* 7, 95–105.

[210] McIntyre, B. D., Riha, S. J., and Ong, C. K., 1997. Competition for water in a hedge-intercrop system. *Field Crops Res.* 52, 151–160.

[211] Baker, E.F.I., 1979. Mixed cropping in northern Nigeria. III. Mixtures of cereals. *Exp. Agric.* 15, 41–48.

[212] Ofori, F., and Stern, W. R., 1987. Cereal-legume intercropping systems. *Adv. Agron.* 41, 41–90.

[213] Hauggaard-Nielsen, H., Jornsgaard, B., Kinane, J., and Jensen, E. S., 2008. Grain legume–cereal intercropping: the practical application of diversity, competition, and facilitation in arable and organic cropping systems. *Renewable Agric.Food Systems* 23, 3–12.

[214] Mathuva, M. N., Rao, M. R., Smithson, P .C., and Coe, R., 1998. Improving maize (*Zea mays*) yields in semiarid highlands of Kenya: agroforestry or inorganic fertilizers? *Field Crops Res.* 55, 57–72.

[215] Hauggaard-Nielsen, H., Ambus, P., and Jensen, E. S., 2003. The comparison of nitrogen use and leaching in sole cropped versus intercropped pea and barley. *Nutr. Cycl. Agroecosys.* 65, 289–300.

[216] Trenbath, B. R., 1974. Biomass productivity of mixtures. *Adv. Agron.* 26, 177–210.

[217] Goodman, D., 1975. The theory of diversity–stability relationships in ecology. *Q. Rev. Biol.* 50, 237–266.

[218] Ives, A. R., and Carpenter, S. R., 2007. Stability and diversity of ecosystems. *Science* 317, 58–62.

[219] Zhu, Y., Chen, H., Fan, J., Wang, Y., et al., 2000. Genetic diversity and disease control in rice. *Nature* 406, 718–722.

[220] Karlen, D. L., Varvel, G. E., Bullock, D. G., and Cruse, R. M., 1994. Crop rotations for the 21st century. *Adv. Agron.* 53, 1–45.

[221] Mitchell, C. C., Arriaga, F. J., Entry, J. A., Novak, J. L., et al., 1996. *The Old Rotation, 1896–1996—100 Years of Sustainable Cropping Research.* Alabama Experiment Station, Auburn, AL.

[222] Bullock, D. G., 1992. Crop rotation. *CRC Crit. Rev. Plant Sci.* 11, 309–326.

[223] Xu, X. M., 2011. A simulation study on managing plant diseases by systematically altering spatial positions of cultivar mixture components between seasons. *Plant Pathol.* 60, 857–865.

[224] Kelly, D., 1994. The evolutionary ecology of mast seeding. *Trends Ecol. Evol.* 9, 465–470.

[225] Heliövaara, K., Väisänen, R., and Simon, C., 1994. Evolutionary ecology of periodical insects. *Trends Ecol. Evol.* 9, 475–480.

[226] Hellman, M. E., 1979. The mathematics of public-key cryptography. *Sci. Am.* 241, 146–157.

[227] Maynard Smith, J., and Price, G. R., 1973. The logic of animal conflict. *Nature* 246, 15–16.

[228] Fisher, R., and Ury, W., 1981. *Getting to Yes.* Houghton Mifflin, New York.

[229] Jennings, P. R., and de Jesus, J., 1968. Studies on competition in rice. I. Competition in mixtures of varieties. *Evolution* 22, 119–124.

[230] Dawkins, R., and Krebs, J. R., 1979. Arms races between and within species. *Proc. Roy. Soc. Lond. B* 205, 489–511.

[231] Zhang, D. Y., Sun, G. J., and Jiang, X. H., 1999. Donald's ideotype and growth redundancy: a game theoretical analysis. *Field Crops Res.* 61, 179–187.

[232] Kokubun, M., 1988. Design and examination of soybean ideotypes. *Japan Agric. Res. Q.* 21, 237–243.

[233] Austin, R. B., Bingham, J., Blackwell, R. D., Evans, L. T., et al., 1980. Genetic improvements in winter wheat yields since 1900 and associated physiological changes. *J. Agric. Sci.* 94, 675–689.

[234] Khush, G., 1999. Green revolution: preparing for the 21st century. *Genome* 42, 646–655.

[235] Lee, E. A., and Tollenaar, M., 2007. Physiological basis of successful breeding strategies for maize grain yield. *Crop Sci.* 47, S202–S215.

[236] Reynolds, M. P., Acevedo, E., Sayre, K. D., and Fischer, R. A., 1994. Yield potential in modern wheat varieties: its association with a less competitive ideotype. *Field Crops Res.* 37, 149–160.

[237] ONeill, P. M., Shanahan, J. F., Schepers, J. S., and Caldwell, B., 2004. Agronomic responses of corn hybrids from different eras to deficit and adequate levels of water and nitrogen. *Agron. J.* 96, 1660–1667.

[238] Campos, H., Cooper, M., Edmeades, G. O., Loffler, C., et al., 2006. Changes in drought tolerance in maize associated with fifty years of breeding for yield in the US corn belt. *Maydica* 51, 369–381.

[239] Farquhar, G. D., Ehleringer, J. R., and Hubick, K. T., 1989. Carbon isotope discrimination and photosynthesis. *Plant Physiol.* 40, 503–537.

[240] Keatinge, J.D.H., and Cooper, P.J.M., 1983. Kabuli chickpea as a winter-sown crop in northern Syria: moisture relations and crop productivity. *J. Agric. Sci.* 100, 667–680.

[241] Denison, R. F., and Loomis, R. S., 1988. *An Integrative Physiological Model of Alfalfa Growth and Development.* University of California, Oakland.

[242] Denison, R. F., Fedders, J. M., and Harter, B. L., 2010. Individual fitness versus whole-crop photosynthesis: solar tracking tradeoffs in alfalfa. *Evol. Appl.* 3, 466–472.

[243] Weiner, J., Andersen, S. B., Wille, W.K.M., Griepentrog, H. W., et al., 2010. Evolutionary agroecology—the potential for cooperative, high-density, weed-suppressing cereals. *Evol. Appl.* 3, 473–479.

[244] Weiner, J., Griepentrog, H. W., and Kristensen, L., 2001. Suppression of weeds by spring wheat *Triticum aestivum* increases with crop density and spatial uniformity. *J. Appl. Ecol.* 38, 784–790.

[245] Page, E. R., Tollenaar, M., Lee, E. A., Lukens, L., et al., 2009. Does the shade avoidance response contribute to the critical period for weed control in maize (*Zea mays*)? *Weed Res.* 49, 563–571.

[246] Henriksson, J., 2001. Differential shading of branches or whole trees: survival, growth, and reproduction. *Oecologia* 126, 482–486.

[247] Sadras, V. O., and Denison, R. F., 2009. Do plant parts compete for resources? An evolutionary viewpoint. *New Phytol.* 183, 565–574.

[248] Novoplansky, A., 1991. Developmental responses of *Portulaca* seedlings to conflicting signals. *Oecologia* 88, 138–140.

[249] Boccalandro, H. E., Ploschuk, E. L., Yanovsky, M. J., Sanchez, R. A., et al., 2003. Increased phytochrome B alleviates density effects on tuber yield of field potato crops. *Plant Physiol.* 133, 1539–1546.

[250] Libenson, S., Rodriguez, V., Pereira, M. L., Sanchez, R. A., et al., 2002. Low red to far-red ratios reaching the stem reduce grain yield in sunflower. *Crop Sci.* 42, 1180–1185.

[251] Childs, D. Z., Metcalf, C.J.E., and Rees, M., 2010. Evolutionary bet-hedging in the real world: empirical evidence and challenges revealed by plants. *Proc. Roy. Soc. Lond. B* 277, 3055–3064.

[252] Simons, A. M., 2011. Modes of response to environmental change and the elusive empirical evidence for bet hedging. *Proc. Roy. Soc. Lond. B* 278, 1601–1609.

[253] Ratcliff, W. C., and Denison, R. F., 2010. Individual-level bet hedging in the bacterium *Sinorhizobium meliloti*. *Curr. Biol.* 20, 1740–1744.

[254] Hamilton, W. D., 1977. The play by nature (review of *The Selfish Gene* by Richard Dawkins). *Science* 196, 757–759.

[255] Muir, W. M., 1996. Group selection for adaptation to multiple-hen cages: selection program and direct responses. *Poult. Sci.* 75, 447–458.

[256] Smith, C. C., and Fretwell, S. D., 1974. The optimal balance between size and number of offspring. *Am. Nat.* 108, 499–506.

[257] Love, O. P., and Williams, T. D., 2008. The adaptive value of stress-induced phenotypes: effects of maternally derived corticosterone on sex-biased investment, cost of reproduction, and maternal fitness. *Am. Nat.* 172, E135–E139.

[258] Swenson, W., Wilson, D. S., and Elias, R., 2000. Artificial ecosystem selection. *Proc. Natl. Acad. Sci. USA* 97, 9110–9114.

[259] Fischer, R. A., 1978. Are your results confounded by intergenotypic competition? In *Proc. 5th International Wheat Genetics Symposium*, vol. II, S. Ramanujam, ed., pp. 767–777. Indian Society of Genetics and Plant Breeding, New Delhi.

[260] Austin, R. B., and Blackwell, R. D., 1980. Edge and neighbour effects in cereal yields trials. *J. Agric. Sci.* 94, 731–734.

[261] Amani, I., Fischer, R. A., and Reynolds, M. P., 1996. Canopy temperature depression association with yield of irrigated spring wheat cultivars in a hot climate. *J. Agron. Crop Sci.* 176, 119–129.

[262] Denison, R. F., and Kiers, E. T., 2011. Life-histories of rhizobia and mycorrhizal fungi. *Curr. Biol.* 21, R775–R785.

[263] Arnold, A. E., Mejia, L. C., Kyllo, D., Rojas, E. I., et al., 2003. Fungal endophytes limit pathogen damage in a tropical tree. *Proc. Natl. Acad. Sci. USA* 100, 15649–15654.

[264] Saikkonen, K., Faeth, S. H., Helander, M., and Sullivan, T. J., 1998. Fungal endophytes: a continuum of interactions with host plants. *Annu. Rev. Ecol. Evol. Syst.* 29, 319–343.

[265] Fribourg, H. A., Chestnut, A. B., Thompson, R. W., McLaren, J. B., et al., 1991. Steer performance in fescue-clover pastures with different levels of endophyte infestation. *Agron. J.* 83, 777–781.

[266] Frederickson, M. E., and Gordon, D. M., 2007. The devil to pay: a cost of mutualism with *Myrmelachista schumanni* ants in "devil's gardens" is increased herbivory on *Duroia hirsuta* trees. *Proc. Roy. Soc. Lond. B* 274, 1117–1123.

[267] Nutman, P. S., 1954. Symbiotic effectiveness in nodulated red clover. I. Variation in host and in bacteria. *Heredity* 8, 35–46.

[268] Erdman, L. W., 1950. Legume inoculation: what it is, what it does. *USDA Farmer's Bull.* 2003, 1–20.

[269] Labandera, C. C., and Vincent, J. M., 1975. Competition between an introduced strain and native Uruguayan strains of *Rhizobium trifolii*. *Plant Soil* 42, 327–347.

[270] Giller, K. E., McGrath, S. P., and Hirsch, P. R., 1989. Absence of nitrogen fixation in clover grown on soil subject to long-term contamination with heavy metals is due to survival of only ineffective *Rhizobium*. *Soil Biol. Biochem.* 6, 841–848.

[271] Moawad, H., Badr El-Din, S.M.S., and Abdel-Aziz, R. A., 1998. Improvement of biological nitrogen fixation in Egyptian winter legumes through better management of *Rhizobium*. *Plant Soil* 204, 95–106.

[272] Denton, M. D., Coventry, D. R., Bellotti, W. D., and Howieson, J. G., 2000. Distribution, abundance and symbiotic effectiveness of *Rhizobium leguminosarum* bv. *trifolii* from alkaline pasture soils in South Australia. *Aust. J. Exp. Agric.* 40, 25–35.

[273] Kiers, E. T., and Denison, R.F., 2008. Sanctions, cooperation, and the stability of plant-rhizosphere mutualisms. *Annu. Rev. Ecol. Evol. Syst.* 39, 215–236.

[274] Klironomos, J. N., 2002. Variation in plant response to native and exotic arbuscular mycorrhizal fungi. *Ecology* 84, 2292–2301.

[275] Johnson, N. C., 1993. Can fertilization of soil select for less mutualistic mycorrhizae? *Ecol. Applic.* 3, 749–757.

[276] Irwin, R. E., Brody, A. K., and Waser, N. M., 2001. The impact of floral larceny on individuals, populations, and communities. *Oecologia* 129, 161–168.

[277] Herre, E. A., and West, S. A., 1997. Conflict of interest in a mutualism: documenting the elusive fig wasp-seed trade-off. *Proc. Roy. Soc. Lond. B* 264, 1501–1507.

[278] Pellmyr, O., and Huth, C. J., 1994. Evolutionary stability of mutualism between yuccas and yucca moths. *Nature* 372, 257–260.

[279] Levey, D. J., Tewksbury, J. J., Cipollini, M. L., and Carlo, T. A., 2006. A field test of the directed deterrence hypothesis in two species of wild chili. *Oecologia* 150, 61–68.

[280] Faeth, S. H., 2002. Are endophytic fungi defensive plant mutualists? *Oikos* 98, 25–36.

[281] Yu, D. W., and Pierce, N. E., 1998. A castration parasite of an ant-plant mutualism. *Proc. Roy. Soc. Lond. B* 265, 375–382.

[282] Stanton, M. L., Palmer, T. M., Young, T. P., Evans, A., et al., 1999. Sterilization and canopy modification of a swollen thorn acacia tree by a plant-ant. *Nature* 401, 578–581.

[283] Hibbett, D. S., Gilbert, L. B., and Donoghue, M. J., 2000. Evolutionary instability of ectomycorrhizal symbioses in basidiomycetes. *Nature* 407, 506–508.

[284] Lutzoni, F., Pagel, M., and Reeb, V., 2001. Major fungal lineages are derived from lichen symbiotic ancestors. *Nature* 411, 937–940.

[285] Douglas, A. E., 2008. Conflict, cheats and the persistence of symbiosis. *New Phytol.* 177, 849–858.

[286] Bethlenfalvay, G. J., Abu-Shakra, S. S., and Phillips, D. A., 1978. Interdependence of nitrogen nutrition and photosynthesis in *Pisum sativum* L. II. Host plant response to nitrogen fixation by *Rhizobium* strains. *Plant Physiol.* 62, 131–133.

[287] Hardin, G., 1968. The tragedy of the commons. *Science* 162, 1243–1248.

[288] Dietz, T., Ostrom, E., and Stern, P. C., 2003. The struggle to govern the commons. *Science* 302, 1907–1912.

[289] Griffin, A. S., and West, S. A., 2002. Kin selection: fact and fiction. *Trends Ecol. Evol.* 17, 15–21.

[290] Simard, S. W., Perry, D. A., Jones, M. D., Myrold, D. D., et al., 1997. Net transfer of carbon between ectomycorrhizal tree species in the field. *Nature* 388, 579–582.

[291] Grafen, A., 1985. A geometric view of relatedness. *Oxf. Surv. Evol. Biol.* 2, 28–89.

[292] Pfeffer, P. E., Douds, D. D., Buecking, H., Schwartz, D. P., et al., 2004. The fungus does not transfer carbon to or between roots in an arbuscular mycorrhizal symbiosis. *New Phytol.* 163, 617–627.

[293] Hagen, M. J., and Hamrick, J. L., 1996. Population level processes in *Rhizobium leguminosarum* bv. *trifolii*: the role of founder effects. *Mol. Ecol.* 5, 707–714.

[294] Dudley, S. A., and File, A. L., 2007. Kin recognition in an annual plant. *Biol. Lett.* 2, 435–438.

[295] Denison, R. F., 2000. Legume sanctions and the evolution of symbiotic cooperation by rhizobia. *Am. Nat.* 156, 567–576.

[296] West, S. A., Kiers, E. T., Simms, E. L., and Denison, R. F., 2002. Sanctions and mutualism stability: why do rhizobia fix nitrogen? *Proc. Roy. Soc. Lond. B* 269, 685–694.

[297] Cowden, C. C., and Peterson, C. J., 2009. A multi-mutualist simulation: applying biological market models to diverse mycorrhizal communities. *Ecol. Model.* 220, 1522–1533.

[298] Kiers, E. T., Rousseau, R. A., West, S. A., and Denison, R. F., 2003. Host sanctions and the legume–rhizobium mutualism. *Nature* 425, 78–81.

[299] Oono, R., Anderson, C. G., and Denison, R. F., 2011. Failure to fix nitrogen by non-reproductive symbiotic rhizobia triggers host sanctions that reduce fitness of their reproductive clonemates. *Proc. Roy. Soc. Lond. B,* 278, 2698–2703.

[300] Kiers, E. T., Duhamel, M., Yugandgar, Y., Mensah, J. A., et al., 2011. Reciprocal rewards stabilize cooperation in the mycorrhizal symbiosis. *Science* 333, 880–882.

[301] Denison, R. F., and Layzell, D. B., 1991. Measurement of legume nodule respiration and $O_2$ permeability by noninvasive spectrophotometry of leghemoglobin. *Plant Physiol.* 96, 137–143.

[302] Friesen, M. L., and Mathias, A., 2010. Mixed infections may promote diversification of mutualistic symbionts: why are there ineffective rhizobia? *J. Evol. Biol.* 23, 323–334.

[303] Singleton, P. W., and Stockinger, K. R., 1983. Compensation against ineffective nodulation in soybean. *Crop Sci.* 23, 69–72.

[304] Simms, E. L., Taylor, D. L., Povich, J., Shefferson, R. P., et al., 2006. An empirical test of partner choice mechanisms in a wild legume-rhizobium interaction. *Proc. Roy. Soc. Lond. B* 273, 77–81.

[305] Kiers, E. T., Rousseau, R. A., and Denison, R. F., 2006. Measured sanctions: legume hosts detect quantitative variation in rhizobium cooperation and punish accordingly. *Evol. Ecol. Res.* 8, 1077–1086.

[306] Kiers, E. T., Hutton, M. G., and Denison, R. F., 2007. Human selection and the relaxation of legume defences against ineffective rhizobia. *Proc. Roy. Soc. Lond. B* 274, 3119–3126.

[307] Denison, R. F., 2007. When can intelligent design of crops by humans outperform natural selection? In *Scale and Complexity in Plant Systems Research: Gene-Plant-Crop Relations*, J.H.J. Spiertz, P. C. Struik, and H. H. van Laar, eds., pp. 287–302. Springer, Dordrecht, The Netherlands.

[308] Johnson, N. C., Copeland, P. J., Crookston, R. K., and Pfleger, F. L., 1992. Mycorrhizae: possible explanation for yield decline with continuous corn and soybean. *Agron. J.* 84, 387–390.

[309] Bever, J. D., 2002. Negative feedback within a mutualism: host-specific growth of mycorrhizal fungi reduces plant benefit. *Proc. Roy. Soc. Lond. B* 269, 2595–2601.

[310] Denison, R. F., Bledsoe, C., Kahn, M. L., O'Gara, F., et al., 2003. Cooperation in the rhizosphere and the "free rider" problem. *Ecology* 84, 838–845.

[311] Jousset, A., Scheu, S., and Bonkowski, M., 2008. Secondary metabolite production facilitates establishment of rhizobacteria by attenuation of protozoan predation and improvement of competitiveness against indigenous microflora. *Funct. Ecol.* 22, 714–719.

[312] English-Loeb, G., Norton, A. P., and Walker, M. A., 2003. Behavioral and population consequences of acarodomatia in grapes on phytoseiid mites (Mesostigmata) and implications for plant breeding. *Entomol. Exp. Appl.* 104, 307–319.

[313] Romero, G. Q., and Benson, W. W., 2005. Biotic interactions of mites, plants, and leaf domatia. *Curr. Opin. Plant Biol.* 8, 436–440.

[314] Romero, G. Q., and Benson, W. W., 2004. Leaf domatia mediate mutualism between mites and a tropical tree. *Oecologia* 140, 609–616.

[315] Norton, A. P., English-Loeb, G., Gadoury, D., and Seem, R. C., 2000. Mycophagous mites and foliar pathogens: leaf domatia mediate tritrophic interactions in grapes. *Ecology* 81, 490–499.

[316] Neve, P., Vila-Aiub, M., and Roux, F., 2009. Evolutionary thinking in agricultural weed management. *New Phytol.* 184, 783–793.

[317] Barrett, S.C.H., 1983. Crop mimicry in weeds. *Econ. Bot.* 37, 255–282.

[318] Schoner, C. A., Norris, R. F., and Chilcote, W., 1978. Yellow foxtail (*Setaria lutescens*) biotype studies: growth and morphological characteristics. *Weed Sci.* 26, 632–636.

[319] Hill, J. E., Smith, R. J., and Bayer, D. E., 1994. Rice weed control: current technology and emerging issues in temperate rice. *Aust. J. Exp. Agric.* 34, 1021–1029.

[320] Losey, J. E., Rayor, L. S., and Carter, M. E., 1999. Transgenic pollen harms monarch larvae. *Nature* 399, 214.

[321] Zangerl, A. R., McKenna, D., Wraight, C. L., Carroll, M., et al., 2001. Effects of exposure to event 176 *Bacillus thuringiensis* corn pollen on Monarch and Black Swallowtail caterpillars under field conditions. *Proc. Natl. Acad. Sci. USA* 98, 11908–11912.

[322] Hodgson, J., 1999. Monarch Bt-corn paper questioned. *Nat. Biotechnol.* 17, 627.

[323] Gould, F., 1998. Sustainability of transgenic insecticidal cultivars: integrating pest genetics and ecology. *Annu. Rev. Entomol.* 43, 701–726.

[324] Alstad, D. N., and Andow, D. A., 1995. Managing the evolution of insect resistance to transgenic plants. *Science* 268, 1894–1896.

[325] Tabashnik, B. E., Gassmann, A. J., Crowder, D. W., and Carriere, Y., 2008. Insect resistance to Bt crops: evidence versus theory. *Nat. Biotechnol.* 26, 199–202.

[326] Fitt, G. P., 2003. Implementation and impact of transgenic Bt cottons in Australia. In *Cotton Production for the New Millennium, Proceedings of the Third World Cotton Research Conference*, A. Swanepoel, ed., pp. 371–381. Agricultural Research Council, Pretoria, South Africa.

[327] Harwood, R. R., 1984. Organic farming research at the Rodale Research Center. In *Organic Farming: Current Technology and Its Role in a Sustainable Agriculture*, D. F. Bezdicek, J. F. Power, D. R. Keeney, and M. J. Wright, eds., pp. 1–17. American Society of Agronomy, Madison, WI.

[328] Martini, E. A., Buyer, J. S., Bryant, D. C., Hartz, T. K., et al., 2004. Yield increases during the organic transition: improving soil quality or increasing experience? *Field Crops Res.* 86, 255–266.

[329] Grafton-Cardwell, E. E., and Gu, P., 2003. Conserving vedalia beetle, *Rodolia cardinalis* (Mulsant) (Coleoptera:Coccinellidae), in citrus: a continuing challenge as new insecticides gain registration. *J. Econ. Entomol.* 96, 1388–1398.

[330] Zalucki, M. P., Adamson, D., and Furlong, M. J., 2009. The future of IPM: whither or wither? *Aust. J. Entomol.* 28, 85–96.

[331] Johnson, M. D., Kellermann, J. L., and Stercho, A. M., 2010. Pest reduction services by birds in shade and sun coffee in Jamaica. *Anim. Conserv.* 13, 140–147.

[332] Gibson, R. W., and Pickett, J. A., 1983. Wild potato repels aphids by release of aphid alarm pheromone. *Nature* 302, 608–609.

[333] Beale, M. H., Birkett, M. A., Bruce, T.J.A., Chamberlain, K., et al., 2006. Aphid alarm pheromone produced by transgenic plants affects aphid and parasitoid behavior. *Proc. Natl. Acad. Sci. USA* 103, 10509–10513.

[334] Cook, S. M., Khan, Z. R., and Pickett, J. A., 2007. The use of push-pull strategies in integrated pest management. *Annu. Rev. Entomol.* 52, 375–400.

[335] Turlings, T.C.J., Loughrin, J. H., McCall, P. J., Rose, U.S.R., et al., 1995. How caterpillar-damaged plants protect themselves by attracting parasitic wasps. *Proc. Natl. Acad. Sci. USA* 92, 4169–4174.

[336] Mlot, C., 2009. Antibiotics in nature: beyond biological warfare. *Science* 324, 1637–1639.

[337] Ratcliff, W. C., and Denison, R. F., 2011. Alternative actions for antibiotics. *Science* 332, 547–548.

[338] Hamilton, W. D., 1971. Geometry for the selfish herd. *J. Theor. Biol.* 31, 295–311.

[339] Diggle, S. P., Gardner, A., West, S. A., and Griffin, A. S., 2007. Evolutionary theory of bacterial quorum sensing: when is a signal not a signal? *Philos. Trans. Roy. Soc. Lond. B* 362, 1241–1249.

[340] Roche, P., Debelle, F., Maillet, F., Lerouge, P., et al., 1991. Molecular basis of symbiotic host specificity in *Rhizobium meliloti*: *nodH* and *nodPQ* genes encode the sulfation of lipo-oligosaccharide signals. *Cell* 67, 1131–1143.

[341] Denison, R. F., and Kiers, E. T., 2004. Lifestyle alternatives for rhizobia: mutualism, parasitism, and forgoing symbiosis. *FEMS Microbiol. Lett.* 237, 187–193.

[342] Fatouros, N. E., Huigens, M. E., van Loon, J.J.A., Dicke, M., et al., 2005. Chemical communication: butterfly anti-aphrodisiac lures parasitic wasps. *Nature* 433, 704.

[343] Bolter, C. J., Dicke, M., van Loon, J.J.A., Visser, J. H., et al., 1997. Attraction of Colorado potato beetle to herbivore-damaged plants during herbivory and after its termination. *J. Chem. Ecol.* 23, 1003–1023.

[344] Brodmann, J., Twele, R., Francke, W., Holzler, G., et al., 2008. Orchids mimic green-leaf volatiles to attract prey-hunting wasps for pollination. *Curr. Biol.* 18, 740–744.

[345] Little, A.E.F., and Currie, C. R., 2008. Black yeast symbionts compromise the efficiency of antibiotic defenses in fungus-growing ants. *Ecology* 89, 1216–1222.

[346] Sen, R., Ishak, H. D., Estrada, D., Dowd, S. E., et al., 2009. Generalized antifungal activity and 454-screening of *Pseudonocardia* and *Amycolatopsis* bacteria in nests of fungus-growing ants. *Proc. Natl. Acad. Sci. USA* 106, 17805–17810.

[347] Fernández-Marín, H., Zimmerman, J. K., Nash, D. R., Boomsma, J. J., et al., 2009. Reduced biological control and enhanced chemical pest management in the evolution of fungus farming in ants. *Proc. Roy. Soc. Lond. B* 276, 2263–2269.

[348] Hughes, W.O.H., Oldroyd, B. P., Beekman, M., and Ratnieks, F.L.W., 2008. Ancestral monogamy shows kin selection is key to the evolution of eusociality. *Science* 320, 1213–1216.

[349] Boomsma, J. J., 2009. Lifetime monogamy and the evolution of eusociality. *Philos. Trans. Roy. Soc. Lond. B* 364, 3191–3207.

[350] West, S. A., and Gardner, A., 2010. Altruism, spite, and greenbeards. *Science* 327, 1341–1344.

[351] Poulsen, M., and Boomsma, J. J., 2005. Mutualistic fungi control crop diversity in fungus-growing ants. *Science* 307, 741–744.

[352] Pinto-Tomas, A. A., Anderson, M. A., Suen, G., Stevenson, D. M., et al., 2009. Symbiotic nitrogen fixation in the fungus gardens of leaf-cutter ants. *Science* 326, 1120–1123.

[353] Lister, R., Gregory, B. D., and Ecker, J. R., 2009. Next is now: new technologies for sequencing of genomes, transcriptomes, and beyond. *Curr. Opin. Plant Biol.* 12, 107–118.

[354] MacLean, D., Jones, J. D., and Studholme, D. J., 2009. Application of "next-generation" sequencing technologies to microbial genetics. *Nat. Rev. Microbiol.* 7, 287–296.

[355] Ricketts, T. H., Daily, G. C., Ehrlich, P. R., and Michener, C. D., 2004. Economic value of tropical forest to coffee production. *Proc. Natl. Acad. Sci. USA* 101, 12579–12582.

[356] McCauley, D. J., 2006. Selling out nature. *Nature* 443, 27–28.

[357] Bains, S., 1996. Sunfish shows the way through the fog. *Science* 272, 653.

[358] Ball, P., 1999. Shark skin and other solutions. *Nature* 400, 507–508.

[359] Gould, F., 1995. Comparisons between resistance management strategies for insects and weeds. *Weed Technol.* 9, 830–839.

[360] Jordan, N. R., and Jannink, J. L., 1997. Assessing the practical importance of weed evolution: a research agenda. *Weed Res.* 37, 237–246.

[361] DeWitt, T. J., and Langerhans, R. B., 2004. Integrated solutions to environmental heterogeneity: theory of multimoment reaction norms. In *Phenotypic*

*Plasticity: Functional and Conceptual Approaches*, T. J. DeWitt and S. M. Scheiner, eds., pp. 98–111. Oxford University Press, New York.

[362] Cheptou, P. O., Carrue, O., Rouifed, S., and Cantarel, A., 2008. Rapid evolution of seed dispersal in an urban environment in the weed *Crepis sancta. Proc. Natl. Acad. Sci. USA* 105, 3796–3799.

[363] Silvertown, J., 1985. When plants play the field. In *Evolution: Essays in Honor of John Maynard Smith*, P. J. Greenwood, P. H. Harvey, and M. Slatkin, eds., pp.143–153. Cambridge University Press, Cambridge, UK.

[364] Philippi, T., and Seger, J., 1989. Hedging one's evolutionary bets, revisited. *Trends Ecol. Evol.* 4, 41–44.

[365] Seavoy, R. E., 1986. *Famine in Peasant Societies*. Greenwood Press, Westport, CT.

[366] Bender, J., 1993. *Future Harvest: Pesticide-Free Farming*. University of Nebraska Press, Lincoln.

[367] Ledford, H., 2010. Plant biologists fear for cress project. *Nature* 464, 154.

[368] Lander, E. S., 2011. Initial impact of the sequencing of the human genome. *Nature* 470, 187–197.

[369] Chen, X., Cui, Z., Vitousek, P. M., Cassman, K. G., et al., 2011. Integrated soil–crop system management for food security. *Proc. Natl. Acad. Sci. USA* 108, 6399–6404.

[370] Sinclair, T. R., and Purcell, L. C., 2005. Is a physiological perspective relevant in a "genocentric" age? *J. Exp. Bot.* 56, 2777–2782.

[371] Bugg, R. L., Sarrantonio, M., Dutcher, J. D., and Phatak, S. C., 1991. Understory cover crops in pecan orchards: possible management systems. *Am. J. Alt. Ag.* 6, 50–62.

[372] Frank, R. H., 2009. The invisible hand, trumped by Darwin? *New York Times*, 12 July 2009.

[373] Kleijn, D., Berendse, F., Smit, R., and Gilissen, N., 2001. Agri-environment schemes do not effectively protect biodiversity in Dutch agricultural landscapes. *Nature* 413, 723–725.

[374] Campbell, D. T., 1969. Reforms as experiments. *Am. Psychol.* 24, 409–429.

[375] Tilman, D., Hill, J., and Lehman, C., 2006. Carbon-negative biofuels from low-input high-diversity grassland biomass. *Science* 314, 1598–1600.

[376] Rupert, M. G., 2008. Decadal-scale changes of nitrate in ground water of the United States, 1988–2004. *J. Environ. Qual.* 37, S-240–S-248.

[377] Rabalais, N. N., Turner, R. E., and Wiseman, W. J., 2002. Gulf of Mexico hypoxia, A.K.A. "The dead zone." *Annu. Rev. Ecol. Evol. Syst.* 33, 235–263.

[378] Fargione, J., Hill, J., Tilman, D., Polasky, S., et al., 2008. Land clearing and the biofuel carbon debt. *Science* 319, 1235–1238.

[379] Angelsen, A., and Kaimowitz, D., 2001. *Agricultural Technologies and Tropical Deforestation*. CABI, Wallingford, UK.

[380] Balmford, A., Green, R. E., and Scharlemann, J.P.W., 2005. Sparing land for nature: exploring the potential impact of changes in agricultural yield on the area needed for crop production. *Global Change Biol.* 11, 1594–1605.

[381] Fischer, J., Brosi, B., Daily, G. C., Ehrlich, P. R., et al., 2008. Should agricultural policies encourage land sparing or wildlife-friendly farming? *Front. Ecol. Environ.* 6, 380–385.

382 Matson, P. A., Naylor, R., and Otiz-Monasterio, I., 1998. Integration of environmental, agronomic, and economic aspects of fertilizer management. *Science* 280, 112–115.

383 Wiesler, F., Bauer, M., Kamh, M., Engels, T., et al., 2002. The crop as indicator for sidedress nitrogen demand in sugar beet production—limitations and perspectives. *J. Plant Nutr.Soil Sci.* 165, 93–99.

384 Allison, P. D., Long, J. S., and Krauze, T. K., 1982. Cumulative advantage and inequality in science. *Am. Sociol. Rev.* 47, 615–625.

385 Wadman, M., 2010. Study says middle sized labs are best. *Nature* 468, 356–357.

386 Deppe, C., 2000. *Breed Your Own Vegetable Varieties: The Gardener's and Farmer's Guide to Plant Breeding and Seed Saving.* Chelsea Green, White River Junction, VT.

# Index

acacia, interactions with ants, 146–147
Africa, 14, 100, 175, 204. *See also* specific
  countries
*Agave*, water-use efficiency of photosynthesis in, 62
age, of earth, 40–41
agricultural engineers, 25
agroecologists, 79–80, 86, 96, 133
agronomy, 4, 20, 203–204
Alakai Swamp, 76, 78
alfalfa (lucerne), 102, 131–135, 157, 166,
  198, 201–202
alarm signals of insect pests, 174–176,
  179, 181
algae, 63, 148
almonds, 50–51, 89
alternate bearing in trees, 51, 117
Altieri, Miguel, 112
amino acids, 30, 38, 58, 95
ancestral-state reconstruction, 38–40, 184
Andow, David, 112
Andrews, T. John, 63
animal agriculture: advantages and disad-
  vantages of, 4, 12, 26, 91–92, 100, 110,
  118, 191–201
animal manure: as resource and pollutant,
  14–15, 18, 25, 101, 106, 202; shortage
  of, 101, 202
Anthony, Susan B., 74
antibiotics: evolution of resistance to,
  36–37; microbial production of, 177,
  ants use of, 182–184
ants: fungus–growing leaf–cutter, 5–6,
  181–185; interactions of trees and,
  146–147
aphids: alarm signals of, 174–176,
  178–179
*Apios americana*, 202
apples, 51, 55
aquifer, 22, 104–105. *See also* groundwater
*Arabidopsis*, 70, 175, 203
archival samples in long-term research,
  19–20
Argentina, 137
Arkansas, 170

asteroid impact risk, 23
Attenborough, David, 140
aurochs, 91
Austin, Roger, 125–126
Australia, 123, 127–130, 136, 170–173

*Bacillus thuringensis* (Bt) bacterial toxin, 68,
  72, 168
back-cross breeding, 54–55, 129
bacteria, 5, 14, 21, 31, 36–37, 142, 161,
  167, 177–178, 182–183, 185. *See
  also* rhizobia
Badger, Murray, 63
Bänziger, Marianne, 65
barley, 17, 111, 198
barnyardgrass, 164–165
Barrett, Spencer, 165
bats to control insect pests, 173, 192
Beachy, Roger, 64
beans, 102, 109–110, 114, 146, 202. *See
  also* soybean
bees, 32, 71, 145, 147, 157, 184, 186
Bell, Mark, 59
Bender, Jim, 200
beneficial predatory insects, 5, 80, 172, 180
Berg, Jeremy, 213
Bergstrom, Carl, 103
bet-hedging, 8, 43–44, 51–52, 138,
  192–194, 197–206, 213
Bever, Jim, 160
Bible, 23
biofilms, 177–178
biofuels, 89, 205, 210
biological control of pests, 80, 172, 180,
  182, 207. *See also* beneficial predatory
  insects
biological nitrogen fixation: 18, 146;
  breeding to improve, 159–160; in fungal
  gardens of ants, 185; by rhizobia maintained by legume sanctions, 156–159.
  *See also* legumes; rhizobia
biotechnology, 3–4, 49, 52–75, 135–138,
  175, 180–181, 191–192, 203–206
birds, 5, 25, 76–77, 85, 103–106, 147, 173,
  208–209; farms as habitat for, 25